Global Warming and the resulting climate change is one of the most serious environmental problems facing the world community. *Global Warming: the Complete Briefing* is the most comprehensive guide available to the subject. A world-renowned expert, Sir John Houghton explores the scientific basis of global warming and the likely impacts of climate change on human society, before addressing the action that could be taken by governments, by industry and by individuals to mitigate the effects. The first edition received excellent reviews, and this completely updated new edition will prove to be the best briefing the student or interested general reader could wish for.

Sir John Houghton CBE, FRS is co-chairman of the Scientific Assessment Working Group of the Intergovernmental Panel on Climate Change and chairman of the UK's Royal Commission on Environmental Pollution. He was Chief Executive of the UK Meteorological Office from 1983 to his retirement in 1991. He is author of *The Physics of Atmospheres* and *The Search for God: Can Science Help?*, and has published numerous research papers and contributed to many influential research documents. Sir John and his wife Sheila live in Wales.

Global Warming:
The Complete Briefing

Global Warming
The Complete Briefing

John Houghton
Co-chairman of the Scientific Assessment Working-Group of the
Intergovernmental Panel on Climate Change

SECOND EDITION

PUBLISHED BY THE PRESS SYNDICATE OF THE UNIVERSITY OF CAMBRIDGE
The Pitt Building, Trumpington Street, Cambridge CB2 1RP, United Kingdom

CAMBRIDGE UNIVERSITY PRESS
The Edinburgh Building, Cambridge CB2 2RU, UK http://www.cup.cam.ac.uk
40 West 20th Street, New York, NY 10011–4211, USA http://www.cup.org
10 Stamford Road, Oakleigh, Melbourne 3166, Australia

First edition © John Houghton 1994
Second edition © Cambridge University Press 1997

First published 1994 by Lion Publishing plc
Second edition published by Cambridge University Press 1997
Reprinted 1998

Printed in the United Kingdom at the University Press, Cambridge

Typeset in Times 10/12 pt [GE]

A catalogue record for this book is available from the British Library

Library of Congress Cataloguing in Publication data

Houghton, John Theodore.
Global warming: the complete briefing/John Houghton.
    p.   cm.
Includes index
ISBN 0 521 62089 9. – ISBN 0 521 62932 2 (pbk.)
I. Global warming. 2. Climate changes. I. Title
QC981.8.G56H68   1997
363.738'74–dc21   97–16353 CIP

ISBN 0 521 62089 9 hardback
ISBN 0 521 62932 2 paperback

Acknowledgments
The following figures and diagrams have been drawn by hardlines, Charlbury: 1.2, 2.1 to 2.3,
2.5 to 2.8, 3.1 3.2a, b, 3.4 to 3.6, 4.3, 4.4, 5.2, 5.3, 5.6, 5.9 to 5.15, 5.17 to 5.19, 6.3, 6.7, 7.4 to
7.10, 8.1, 8.2, 9.1, 9.3, 9.4, 11.1 to 11.4 and 11.6 to 11.15.

To my grandchildren, Daniel, Hannah, Esther, Maxwell and Jonathan and their generation

Contents

Introduction to the First Edition

Climate change and global warming are well up on the current political agenda. There are urgent questions everyone is asking: are human activities altering the climate? Is global warming a reality? How big are the changes likely to be? Will there be more serious disasters; will they be more frequent? Can we adapt to climate change or can we change the way we do things so that we can slow down the change or even prevent it occurring?

Because the Earth's climate system is highly complex, and because human behaviour and reaction to change is even more complex, providing answers to these questions is an enormous challenge to the world's scientists. As with many scientific problems only partial answers are available, but our knowledge is evolving rapidly, and the world's scientists have been addressing the problems with much energy and determination.

Three major pollution issues are often put together in people's minds: global warming, ozone depletion (the ozone hole) and acid rain. Although there are links between the science of these three issues (the chemicals which deplete ozone and the particles which are involved in the formation of acid rain also contribute to global warming), they are essentially three distinct problems. Their most important common feature is their large scale. In the case of acid rain the emissions of sulphur dioxide from one nation's territory can seriously affect the forests and the lakes of countries which may be downwind of the pollution. Global warming and ozone depletion are examples of global pollution – pollution in which the activities of one person or one nation can affect all people and all nations. It is only during the last thirty years or so that human activities have been of such a kind or on a sufficiently large scale that their effects can be significant globally. And because the problems are global, all nations have to be involved in their solution.

The key intergovernmental body which has been set up to assess the problem of global warming is the Intergovernmental Panel on Climate Change (IPCC), formed in 1988. At its first meeting in November of that year in Geneva, the Panel's first action was to ask for a scientific report so that, so far as they were known, the scientific facts about global warming could be established. It was imperative that politicians were given a solid scientific base from which to develop the requirements for action.

That first scientific report was published at the end of May 1990. On Monday 17 May I presented a preview of it to the then British Prime Minister, Mrs Margaret Thatcher, and members of her Cabinet at 10, Downing Street in London. I had been led to expect many interruptions and questions during my presentation. But the thirty or so Cabinet members and officials in the historic Cabinet room heard me in silence. They were clearly very interested in the report, and the questions and discussion afterwards demonstrated a large degree of concern for the world's environmental problems.

Since then the interest of many political leaders has been aroused – as has been shown by their attendance at two important world conferences concerned with global warming: the Second World Climate Conference in Geneva in 1990 and the United Nations Conference on Environment and Development (UNCED) in Rio de Janeiro in 1992. The Rio conference with over 25,000 people attending the main sessions and the many side meetings, was the largest conference ever held. Never before had a single conference seen so many of the world's leaders, and for that reason it is often referred to as the Earth Summit.

Much of the continuing assessment of climate change has been focused on the IPCC and its three working groups dealing respectively with science, impacts and response strategies. The IPCC's first report published in 1990 was a key input to the international negotiations which prepared the agenda for the UNCED Conference in Rio de Janeiro; it was that IPCC assessment which provided much of the impetus for the Framework Convention on climate change signed at Rio by over 160 countries. As chairman or co-chairman of the Science Working Group I have been privileged to work closely with hundreds of scientific colleagues in many countries who readily gave of their time and expertise to contribute to the IPCC work.

For this book I have drawn heavily on the 1990 and 1992 reports of all three working groups of IPCC. Further, in putting forward options for action I have followed the logic of the Climate Convention. What I have said I believe to be consistent with the IPCC reports and with the implications of the Climate Convention. However, I must also emphasize that the choice of material and any particular views I put forward are entirely my own and should in no way be construed as the views of the IPCC.

During the preparation of both IPCC reports so far there has been considerable scientific debate about just how much can be said about likely climate change next century. Some researchers initially felt that the uncertainties were such that scientists should refrain from making any estimates or predictions for the future. However, it soon became clear that scientists have a responsibility to communicate the best possible information about the likely magnitude of climate change, along with clear statements of the assumptions made and the level of uncertainty in the estimates. Like weather forecasters, their results will not be entirely accurate, but can provide useful guidance.

Many books have been published on global warming. This book differs from the others because I have attempted to describe the science of global warming, its impacts and what action might be taken in a way which the intelligent non-scientist can understand. Although there are many numbers in the book – I believe the quantification of the problem to be very important – there are no mathematical equations. I have also used the minimum of jargon in the main text. Some technical explanations which would be of interest to the scientifically trained are included in some of the boxes. Others contain further material of specific interest.

I am grateful to many who have helped me with the provision and preparation of particular material for this book and to those who have read and helpfully commented on my drafts. There have been those who have been involved with the IPCC: Bert Bolin, the IPCC Chairman, Gylvan Meira Filho, my co-chairman on the IPCC Science Working Group, Robert Watson, co-chairman of the IPCC Working Group on Impacts and Response Strategies, Bruce Callander, Chris Folland, Neil Harris, Katherine Maskell, John Mitchell, Martin Parry, Peter Rowntree, Catherine Senior and Tom Wigley. Others I wish to thank are Myles Allen, David Carson, Jonathan Gregory, Donald Hay, David Fisk, Kathryn Francis, Michael Jefferson, Geoffrey Lean and John Twidell. The staff at Lion Publishing, Rebecca Winter, Nicholas Rous and Sarah Hall, have been most helpful in preparing the book for publication, especially in ensuring that it is as attractive and readable as possible. Finally, I owe an especial debt to my wife, Sheila, who gave me strong encouragement to write the book in the first place, and who has continued her encouragement and support through the long hours of its production.

Introduction to the Second Edition

Since the publication of the first edition nearly three years ago, interest in the issue of Global Warming and concern about it has continued to grow. The Framework Convention on Climate Change (FCCC) agreed at the Earth Summit in 1992 has been ratified and machinery for its implementation is gradually being developed. At the end of 1995, the IPCC produced a further comprehensive report updating the 1990 report. Although the main conclusions have not changed, much has been added to the detail of our knowledge regarding all aspects of the issue, the science, the impacts and the possible response. This revised edition takes into account this further information from the 1995 IPCC reports.

In the first edition I included a chapter, chapter 8, with the heading 'Why should we be concerned?' which addresses the question of the responsibility of humans for the Earth and for looking after the environment. In it I presented something of the basis for my personal motivation as a Christian for being concerned with environmental problems. Although I believe that it important that science is presented in the broad context of human values, I realised that the inclusion of such a chapter was something of a departure and wondered how it would be received.

Some have expressed surprise that in the middle of a science book, there should be, unusually, a chapter of this kind which deals with ethical and religious issues. However, it has been pleasing that scientific colleagues and reviewers of the book have referred favourably to the chapter stressing the value and importance of placing environmental science in the context of the reasons for its pursuit. For instance, John Perry, in the *Bulletin of the American Meteorological Society*, writes:

Many scientists, including avowed agnostics such as myself, will find this forthright declaration of religious belief and divine purpose a bit startling in an otherwise rigorously scientific volume. However, in a line of argument that I have no difficulty whatever in supporting, Houghton demonstrates that the domains of science and religion are simply complementary ways of looking at truth. The former deals with how the world works and the latter with why. In Houghton's framework, we and the earth are each other's reasons for existence in a divine plan that we must struggle to understand but must inescapably follow. Thus, Houghton holds that we have no choice but to care for the earth solicitously as its 'gardeners' in a 'partnership with God'. His lucid precis of the complex factual substance of global warming is an authoritative guide to the issue's scientific dimensions; his inspiring synthesis of science, faith and stewardship is an even more illuminating handbook to its moral and ethical dimensions. Together, they constitute a uniquely valuable Baedeker to one of the most important issues of our science and our time.

In revising chapter 8 for this edition, I have been somewhat more objective and less personal – which I felt was more appropriate for student readers from a wide range of disciplines, for whom the edition is particularly suited. As a didactic aid I have also included a number of problems and questions for discussion at the end of all the chapters.

Some of my colleagues sometimes comment on how formidable is the task of stewardship of the Earth feeling that it is perhaps beyond the capability of the human race to tackle it adequately. I feel optimistic about it, however, for three main reasons. Firstly, I have seen how the world's scientists, coming from very different countries, cultures and backgrounds, have worked closely and responsibly in the IPCC to provide a consensus presentation of the science of global warming. Secondly, the technologies required to provide for greater efficiency in the use of fossil fuels and for their replacement with renewable sources of energy are available and, when developed on the necessary scale, also affordable. Thirdly, my belief in God's commitment to the material world coupled with his offer of partnership in caring for it, makes stewardship of the Earth an especially exciting and challenging activity.

In the preparation of this revised volume I wish to express again my gratitude to the scientific colleagues with whom I have worked in the ongoing activity of the IPCC and from whom I have learnt much. My thanks are also due to John Twidell and Michael Banner who have commented on particular chapters, and to Catherine Flack, Matt Lloyd and other staff of the Cambridge University Press for their competence, courtesy and assistance in the preparation of the book.

John Houghton
1997

1 Global Warming and Climate Change

The phrase 'global warming' has become familiar to many people as one of the important environmental issues of our day. Many opinions have been expressed concerning it, from the doom-laden to the dismissive. This book aims to state the current scientific position on global warming clearly, so that we can make informed decisions on the facts.

Is the climate changing?

In the year 2060 my grandchildren will be approaching seventy; what will their world be like? Indeed, what will it be like during the seventy years or so of their normal life span? Many new things have happened in the last seventy years which could not have been predicted in the 1920s. The pace of change is such that even more novelty can be expected in the next seventy. It is fairly certain that the world will be even more crowded and more connected. Will the increasing scale of human activities affect the environment? In particular,will the world be warmer? How is its climate likely to change?

Before studying future climate changes, what can be said about climate changes in the past? In the more distant past there have been very large changes. The last million years have seen a succession of major ice ages interspersed with warmer periods. The last of these ice ages began to come to an end about 20,000 years ago and we are now in what is called an interglacial period. Chapter 4 will focus on these times far back in the past. But have there been changes in the very much shorter period of living memory – over the past few decades?

Variations in day-to-day weather are occurring all the time; they are very much part of our lives. The climate of a region is its average weather over a period which may be a few months, a season or a few years. Variations in climate are also very familiar to us. We describe summers as wet or dry, winters as mild, cold or stormy. In the British Isles, as in many parts of the world, no season is the same as the last or indeed the same as any previous season, nor will it be repeated in detail next time round. Most of these variations we take for granted; they add a lot of interest to our lives. Those we particularly notice are the extreme situations and the climate disasters (for instance, Fig. 1.1 shows the significant climate events and disasters during the year 1991). Most of the worst disasters in the world are, in fact, weather- or climate-related. Table 1.1 lists them in order of severity although it does not include droughts, whose effects occur more slowly, but which are probably the most damaging disasters of all.

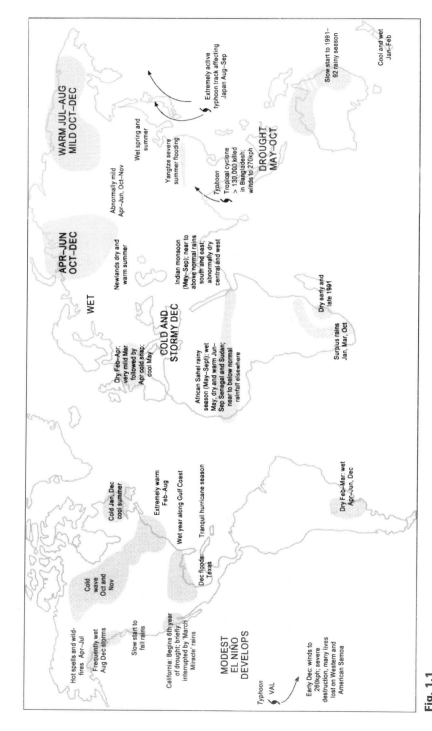

Fig. 1.1

Significant climate anomalies and events during 1991 as recorded by the Climate Analysis Centre of the United States[1].

Table 1.1
Natural disasters 1947–1980 in order of severity[2].

Type of disaster	Deaths
1. Tropical cyclones, hurricanes, typhoons	499,000
2. Earthquakes	450,000
3. Floods (other than associated with 1)	194,000
4. Tornadoes and thunderstorms	29,000
5. Snowstorms	10,000
6. Volcanoes	9,000
7. Heatwaves	7,000
8. Avalanches	5,000
9. Landslides	5,000
10. Tidal waves (Tsunamis)	5,000

The 1980s: a remarkable decade

The 1980s were unusually warm. Globally speaking, the decade has been the warmest since accurate records began somewhat over a hundred years ago and unusually warm years have continued into the 1990s. In terms of global average near surface air temperature, the year 1995 was the warmest on record and eight of the nine warmest years in the record occurred in the 1980s and early 1990s.

The period has also been remarkable (just how remarkable will be considered later) for the frequency and intensity of extremes of weather and climate. For example, periods of unusually strong winds have been experienced in western Europe. During the early hours of the morning of 16 October 1987, over fifteen million trees were blown down in south-east England and the London area. The storm also hit Northern France, Belgium and Holland with ferocious intensity; it turned out to be the worst storm experienced in the area since 1703. Storm-force winds of similar intensity but covering a greater area of western Europe struck on several occasions in January and February 1990.

But those storms in Europe were mild by comparison with the much more intense and damaging storms other parts of the world have experienced during these years. About eighty hurricanes and typhoons – other names for tropical cyclones – occur around the tropical oceans each year, familiar enough to begiven names. Hurricane Gilbert, which caused devastation on the island of Jamaica and the coast of Mexico in 1988, and Hurricane Andrew, which caused a great deal of damage in Florida and other regions of the southern United States in 1992, have been notable recent examples. Low-lying areas such as Bangladesh are particularly vulnerable to the storm surges associated with tropical cyclones; the combined effect of intensely low atmospheric pressure, extremely strong winds and high tides causes a surge of water which can reach far inland. In one of the worst such disasters this century over 250,000 people were drowned in Bangladesh in 1970. The people of that country experienced another storm of similar proportions in 1991 and smaller surges are a regular occurrence there.

Table 1.2
Losses (in thousand millions of US dollars adjusted to 1992 prices) in major windstorm catastrophes 1960–92 (mostly in North America and Europe), estimated by a research group advising the insurance industry[3].

	Decade 1960–69	Decade 1970–79	Decade 1980–89	10 years 1983–92
Number of windstorm catastrophes	8	14	29	31
Economic losses	23	34	38	88
Insured losses	5	8	19	52

The increase in storm intensity during recent years has been tracked by the insurance industry, which has been hit hard by recent disasters. Until the mid-1980s, it was widely thought that windstorms with insured losses exceeding one thousand million US dollars were only possible, if at all, in the United States. But the gales that hit western Europe in October 1987 heralded a series of windstorm disasters which make losses of 10 thousand million dollars seem commonplace. Hurricane Andrew, for instance, left in its wake insured losses estimated at 16 thousand million dollars. The estimates in Table 1.2 illustrate how the numbers and extent of such disasters have increased during the past three decades. The rate of economic loss has risen by a factor of four since the 1960s while the increase in insured losses is almost tenfold. Although some of this increase is due to the growth in population over this period in particularly vulnerable areas, a significant part of it seems to have arisen from the increased storminess in the late 1980s and early 1990s.

Windstorms are by no means the only weather and climate extremes that cause disasters. Floods due to unusually intense or prolonged rainfall or droughts because of long periods of reduced rainfall (or its complete absence) can be even more devastating to human life and property. These events occur frequently in many parts of the world especially in the tropics and sub-tropics. There have been notable examples during the last decade. In 1988, the highest flood levels ever recorded occurred in Bangladesh; 80 per cent of the entire country was affected. The Yangtze river region of China experienced a devastating flood in 1991. In 1993, flood waters rose to levels higher than ever recorded in the region of the Mississippi and Missouri rivers in the United States, flooding an area equivalent in size to one of the Great Lakes. Large areas of Africa, both north and south, and of Australia have had droughts on a scale and for longer periods than any in living memory.

Because of the likely locations of floods and droughts, they often bear most heavily on the most vulnerable in the world, who can have little resilience to major disasters. Figure 1.2 shows that climate-related disasters account for more than half of all disasters for the continent of Africa and illustrates the scale of the problem.

The El Niño event

Rainfall patterns which lead to floods and droughts in tropical and semi-tropical areas are strongly influenced by the surface temperature of the oceans

Fig. 1.2
Recorded disasters in
Africa, 1980–1989,
estimated by the
Organization for African
Unity[4].

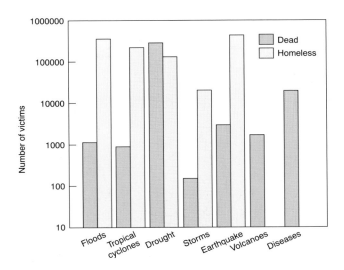

around the world, particularly the pattern of ocean surface temperature in the Pacific off the coast of South America (see Chapter 5). About every three to five years a large area of warmer water appears and persists for a year or more. Because they usually occur around Christmas these are known as El Niño ('the boy child') events. They have been well known for centuries to the countries along the coast of South America because of their devastating effect on the fishing industry; the warm top waters of the ocean prevent the nutrients from lower, colder levels required by the fish from reaching the surface.

A particularly intense El Niño occurred in 1982–83; the anomalous highs in ocean surface temperature compared to the average reached 7 °C. Droughts and floods somewhere in almost all the continents were associated with that El Niño (Fig. 1.3). Like many events associated with weather and climate, El Niños often differ very much in their detailed character. For instance, the El Niño event which began in 1990 and reached maturity early in 1992, apart from some weakening in mid-1992, continued to be dominated by the warm phase until 1995. The exceptional floods in the central United States and in the Andes, and the droughts in Australia and Africa, are probably linked with this unusually protracted El Niño. Studies with computer models of the kind described later in Chapter 5 provide a scientific basis for the links between the El Niño and these extreme weather events; they also give some confidence that useful forecasts of such disasters may one day be possible.

The effect of volcanic eruptions on temperature extremes

Volcanoes inject enormous quantities of dust and gases into the upper atmosphere. Large amounts of sulphur dioxide are included, which through photochemical reactions using the sun's energy are transformed to sulphuric acid and sulphate particles. Typically these particles remain in the stratosphere (the

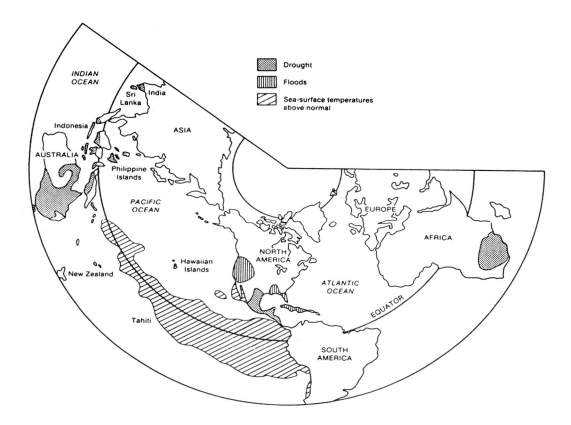

Fig. 1.3
Regions where
droughts and floods
occurred associated
with the 1982–83 El
Niño[5].

region of atmosphere above about 10 km in altitude) for several years before they fall into the lower atmosphere and are quickly washed out by rainfall. During this period they disperse around the whole globe and cut out some of the radiation from the sun, thus tending to cool the lower atmosphere.

One of the largest volcanic eruptions this century was that from Mount Pinatubo in the Philippines on 12 June 1991 which injected about 20 million tonnes of sulphur dioxide into the stratosphere together with enormous amounts of dust. This stratospheric dust caused spectacular sunsets around the world for many months following the eruption. The amount of radiation from the sun reaching the lower atmosphere fell by about 2 per cent. Global average temperatures lower by about a quarter of a degree Celsius were experienced for the following two years. There is also evidence that some of the unusual weather patterns of 1991 and 1992, for instance unusually cold winters in the Middle East and mild winters in western Europe, were linked with effects of the volcanic dust.

Vulnerable to change

Over the centuries different human communities have adapted to their particular climate; any large change to the average climate tends to bring stress of one kind or another. It is particularly the extreme climate events and climate disasters which emphasize the importance of climate to our lives and which demonstrate to countries around the world their vulnerability to climate change – a vulnerability which is enhanced by rapidly increasing demands on resources.

But the question must be asked: how remarkable are these events? Do they point to a changing climate due to human activities? Do they provide evidence for global warming because of the increased carbon dioxide and other greenhouse gases being emitted into the atmosphere by burning fossil fuels?

Here a note of caution must be sounded. The range of normal natural climate variation is large. Climate extremes are nothing new. Climate records are continually being broken. In fact, a month without a broken record somewhere would itself be something of a record! Changes in climate which indicate a genuine long-term trend can only be identified after many years.

However, we know for sure that, because of human activities especially the burning of fossil fuels, carbon dioxide in the atmosphere has been increasing over the past two hundred years and more substantially over the past fifty years. To identify climate change related to this carbon dioxide increase, we need to look for trends in global warming over similar lengths of time. They are long compared with both the memories of a generation and the period for which accurate and detailed records exist. Although, therefore, it can be ascertained that there has been more storminess, for instance, in the region of the north Atlantic during the late 1980s and early 1990s than there was in the previous two decades, it is not clear whether those years were that exceptional compared with other periods in the previous hundred years. There is even more difficulty in tracking detailed climate trends in many other parts of the world, owing to the lack of adequate records; further, trends in the frequency of rare events are very difficult to detect.

The generally cold period worldwide during the 1960s and early 1970s caused speculation that the world was heading for an ice age. A British television programme about climate change called 'The ice age cometh' was prepared in the early 1970s and widely screened – but the cold trend soon came to an end. We must not be misled by our relatively short memories.

What is important is continually to make careful comparisons between practical observations of the climate and its changes and what scientific knowledge leads us to expect. During the last few years, as the occurrence of extreme events has made the public much more aware of environmental issues, scientists in their turn have become somewhat more sure about just what human activities are doing to the climate. Later chapters will look in detail at the science of global warming and at the climate changes that we can expect, as well as investigating how these changes fit in with the recent climate record. Here, however, is a brief outline of current scientific thinking on the problem.

The problem of global warming

Human industry and other activities such as deforestation are emitting increasing quantities of gases, in particular the gas carbon dioxide, into the atmosphere. Every year these emissions currently add to the carbon already present in atmospheric carbon dioxide a further seven thousand million tonnes, much of which is likely to remain there for a period of a hundred years or more. Because carbon dioxide is a good absorber of heat radiation coming from the Earth's surface, increased carbon dioxide acts like a blanket over the surface, keeping it warmer than it would otherwise be. With the increased temperature the amount of water vapour in the atmosphere also increases, providing more of a blanket effect and causing it to be even warmer.

Being kept warmer may sound appealing to those of us who live in cool climates. However, an increase in global temperature will lead to global climate change. If the change were small and occurred slowly enough we would almost certainly be able to adapt to it. However, with the rapid expansion taking place in the world's industry the change is unlikely to be either small or slow. The estimate I present in later chapters is that, in the absence of efforts to curb the rise in the emissions of carbon dioxide, the global average temperature will rise by about a quarter of a degree Celsius every ten years – or about two and a half degrees in a century.

This may not sound very much, especially when it is compared with normal temperature variations from day to night or between one day and the next. But it is not the temperature at one place but the temperature averaged over the whole globe. The predicted rate of change of two and a half degrees a century is probably faster than the global average temperature has changed at any time over the past ten thousand years. And as there is a difference in global average temperature of only about five or six degrees between the coldest part of an ice age and the warm period in between ice ages (see Fig. 4.4), we can see that a few degrees in this global average can represent a big change in climate.

Not all the climate changes will in the end be adverse. While some parts of the world experience more frequent or more severe droughts or floods, other parts perhaps in the sub-arctic may become more habitable. Even there, though, the likely rate of change will cause problems: large damage to buildings will occur in regions of melting permafrost, and trees in sub-arctic forests like trees elsewhere will need time to adapt to new climatic regimes.

Scientists are confident about the fact of global warming and climate change due to human activities. However, substantial uncertainty remains about just how large the warming will be and what will be the patterns of change in different parts of the world. Although some indications can be given, scientists cannot yet say with a lot of detail which regions will be most affected and in what way. Intensive research is needed to improve the confidence in scientific predictions.

Uncertainty and response

Until the predictions improve to the point where they can be used as a clear guide to action, politicians and others making decisions are faced with the need to weigh scientific uncertainty against the cost of the various actions which could be taken in response to the threat of climate change. Some action can be taken easily at relatively little cost (or even at a net saving of cost), for instance the development of programmes to conserve and save energy, and many schemes for reducing deforestation and encouraging the planting of trees. Other actions such as a large shift to energy sources which do not lead to significant carbon dioxide emissions (for example, renewable sources – biomass, hydro, wind or solar energy) in both the developed and the developing countries of the world will take some time. But here, too, a start can be made. What is important is that plans are made now in preparation for the major changes that will almost certainly be required.

In the following chapters I shall first explain the science of global warming, the evidence for it and the current state of the art regarding climate prediction. I shall then go on to say what is known about the likely impacts of climate change on human life – on water and food supplies for instance. The questions of why we should be concerned for the environment and what action should be taken in the face of scientific uncertainty is followed by consideration of the technical possibilities for large reductions in the emissions of carbon dioxide and how these might affect our energy sources and usage, including means of transport.

Finally I will address the issue of the 'global village'. So far as the environment is concerned, national boundaries are becoming less and less important; pollution in one country can now affect the whole world. And it is clear from our current scientific understanding that global warming poses a global challenge, which must be met by global solutions.

Questions   **1**   Look through recent copies of newspapers and magazines for articles which mention climate change, global warming or the greenhouse effect. How many of the statements made are accurate?

2   Make up a simple questionnaire about climate change, global warming and the greenhouse effect to find out how much people know about these subjects, their relevance and importance. Analyse results from responses to the questionnaire in terms of the background of the respondents. Suggest ways in which people could be better informed.

Notes

1 From *World Climate News, No. 1*, June 1992: World Meteorological Organization, Geneva.
2 After B. V. Shah, 1983, quoted. in 'Natural Disaster Reduction: how meteorological services can help', *WMO, No. 722*, 1989, World Meteorological Organization, Geneva.
3 From G. Berz and K. Conrad, 'Winds of change', *The Review*, June 1993, pp. 32–5.
4 From 'The role of the World Meteorological Organization in the International Decade for Natural Disaster Reduction' *WMO, No. 745*, 1990, World Meteorological Organization, Geneva.
5 Adapted from T. Y. Canby, 'El Niño's ill wind', *Natn. Geogr. Mag.*, 1984, pp. 144–83.

2 The Greenhouse Effect

The basic principle of global warming can be understood by considering the radiation energy from the sun which warms the Earth's surface and the thermal radiation from the Earth and the atmosphere which is radiated out to space. On average these two radiation streams must balance. If the balance is disturbed (for instance by an increase in atmospheric carbon dioxide) it can be restored by an increase in the Earth's surface temperature.

How the Earth keeps warm

To explain the processes which warm the Earth and its atmosphere, I will begin with a very simplified Earth. Suppose we could, all of a sudden, remove from the atmosphere all the clouds, the water vapour, the carbon dioxide and all the other minor gases and the dust leaving an atmosphere of nitrogen and oxygen only. Everything else remains the same. What, under these conditions, would happen to the atmospheric temperature?

The calculation is an easy one, involving a relatively simple radiation balance. Radiant energy from the sun falls on a surface of one square metre in area outside the atmosphere and directly facing the sun at a rate of about 1370 watts – about the power radiated by a reasonably sized domestic electric fire. However, few parts of the Earth's surface face the sun directly and in any case for half the time they are pointing away from the sun at night, so that the average energy falling on one square metre of a level surface outside the atmosphere is only one quarter of this[1] or about 343 watts. As this radiation passes through the atmosphere a small amount, about 6 per cent, is scattered back to space by atmospheric molecules. About 10 per cent on average is reflected back to space from the land and ocean surface. The remaining 84 per cent, or about 288 watts per square metre on average, remains actually to heat the surface – the power used by three good-sized incandescent electric light bulbs.

To balance this incoming energy, the Earth itself must radiate on average the same amount of energy back to space (Fig. 2.1) in the form of thermal radiation. All objects emit this kind of radiation; if they are hot enough we can see the radiation they emit. The sun at a temperature of about 6,000 °C looks white; an electric fire at 800 °C looks red. Cooler objects emit radiation which cannot be seen by our eyes and which lies at wavelengths beyond the red end of the spectrum – infrared radiation (sometimes called long-wave radiation to distinguish it from the short-wave radiation from the sun). On a clear, starry winter's night we are very aware of the cooling effect of this kind of radiation being emitted by the Earth's surface into space – it often leads to the formation of frost.

Table 2.1

The composition of the atmosphere, the main constituents (nitrogen and oxygen) and the greenhouse gases as in 1995.

Gas	Concentration: fraction* or parts per million by volume (ppmv)
Nitrogen (N_2)	0.78*
Oxygen (O_2)	0.21*
Water vapour (H_2O)	variable (0–0.02*)
Carbon dioxide (CO_2)	360
Methane (CH_4)	1.8
Nitrous Oxide (N_2O)	0.3
CFCs	0.001
Ozone (O_3)	variable (0–1,000)

Fig. 2.1
The radiation balance of planet Earth. The net incoming solar radiation is balanced by outgoing thermal radiation from the Earth.

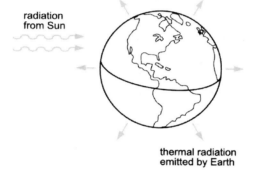

radiation from Sun

thermal radiation emitted by Earth

The amount of thermal radiation emitted by the Earth's surface depends on its temperature – the warmer it is, the more radiation is emitted. The amount of radiation also depends on how absorbing the surface is; the greater the absorption, the more the radiation. Most of the surfaces on the Earth, including ice and snow, would appear 'black' if we could see them at infrared wavelengths; that means that they absorb nearly all the thermal radiation which falls on them instead of reflecting it. It can be calculated[2] that, to balance the energy coming in, the average temperature of the Earth's surface must be –6 °C to radiate the right amount[3]. This is much colder than is actually the case. In fact, an average of temperatures measured near the surface all over the Earth – over the oceans as well as over the land – averaging, too, over the whole year, comes to about 15 °C. Some factor not yet taken into account is needed to explain this discrepancy.

The greenhouse effect

The gases nitrogen and oxygen which make up the bulk of the atmosphere (Table 2.1 gives details of the atmosphere's composition), neither absorb nor emit thermal radiation. It is the water vapour, carbon dioxide and some other minor gases present in the atmosphere in much smaller quantities (Table 2.1) which absorb some of the thermal radiation leaving the surface, acting as a partial blanket for this radiation and causing the difference of 21 °C or so between the actual average surface temperature on the Earth of about 15 °C and the figure of −6 °C which applies when the atmosphere contains nitrogen and oxygen only[4]. This blanketing is known as the *natural greenhouse effect* and the gases are known as greenhouse gases. It is called 'natural' because all the atmospheric gases (apart from the chlorofluorocarbons – CFCs) were there long before human beings came on the scene. Later on I will mention the *enhanced greenhouse effect*: the added effect caused by the gases present in the atmosphere due to human activities such as the burning of fossil fuels and deforestation.

The basic science of the greenhouse effect has been known since early in the 19th century (see box) when the similarity between the radiative properties of the Earth's atmosphere and of the glass in a greenhouse (Fig 2.2) was first pointed out – hence the name 'greenhouse effect'. In a greenhouse, visible radiation from the sun passes almost unimpeded through the glass and is absorbed by the plants and the soil inside. The thermal radiation which is emitted

---

Pioneers of the science of the greenhouse effect[5]

The warming effect of the greenhouse gases in the atmosphere was first recognized in 1827 by the French scientist Jean-Baptiste Fourier, best known for his contributions to mathematics. He also pointed out the similarity between what happens in the atmosphere and in the glass of a greenhouse, which led to the name 'greenhouse effect'. The next step was taken by a British scientist, John Tyndall, who, around 1860, measured the absorption of infrared radiation by carbon dioxide and water vapour; he also suggested that a cause of the ice ages might be a decrease in the greenhouse effect of carbon dioxide. It was a Swedish chemist, Svante Arrhenius, in 1896, who calculated the effect of an increasing concentration of greenhouse gases; he estimated that doubling the concentration of carbon dioxide would increase the global average temperature by 5 to 6 °C, an estimate not too far from our present understanding. Nearly fifty years later, around 1940, G. S. Callendar, working in England, was the first to calculate the warming due to the increasing carbon dioxide from the burning of fossil fuels.

The first expression of concern about the climate change which might be brought about by increasing greenhouse gases was in 1957, when Roger Revelle and Hans Suess of the Scripps Institute of Oceanography in California published a paper which pointed out that in the build-up of carbon dioxide in the atmosphere, human beings are carrying out a large-scale geophysical experiment. In the same year, routine measurements of carbon dioxide were started from the observatory on Mauna Kea in Hawaii. The rapidly increasing use of fossil fuels since then, together with growing interest in the environment, has led to the topic of global warming moving up the political agenda through the 1980s, and eventually to the Climate Convention signed in 1992 – of which more in later chapters.

Fig. 2.2
A greenhouse has a similar effect to the atmosphere on the incoming solar radiation and the emitted thermal radiation.

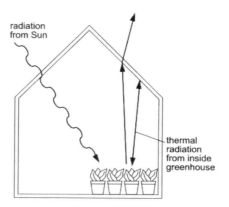

by the plants and soil is, however, absorbed by the glass which re-emits some of it back into the greenhouse. The glass thus acts as a 'radiation blanket' helping to keep the greenhouse warm.

However, the transfer of radiation is only one of the ways heat is moved around in a greenhouse. A more important means of heat transfer is due to convection in which less dense warm air moves upwards and more dense cold air moves downwards. A familiar example of this process is the use of convective electric heaters in the home which heat a room by stimulating convection in it. The situation in the greenhouse is therefore more complicated than would be the case if radiation was the only process of heat transfer.

Mixing and convection are also present in the atmosphere, although on a much larger scale, and in order to achieve a proper understanding of the greenhouse effect, convective heat transfer processes in the atmosphere must be taken into account as well as radiative ones.

Within the atmosphere itself (at least in the lowest three-quarters or so of the atmosphere up to a height of about 10 km which is called the troposphere) convection is, in fact, the dominant process for transferring heat. It acts as follows. The surface of the Earth is warmed by the sunlight it absorbs. Air close to the surface is heated and rises because of its lower density. As the air rises it expands and cools – just as the air cools as it comes out of the valve of a tyre. As some air masses rise, other air masses descend, so the air is continually turning over as different movements balance each other out – a situation of convective equilibrium. Temperature in the troposphere falls with height at a rate determined by these convective processes; the drop turns out on average to be about 6 °C per kilometre of height (Fig. 2.3).

A picture of the transfer of radiation in the atmosphere may be obtained by looking at the thermal radiation emitted by the Earth and its atmosphere as observed from instruments on satellites orbiting the Earth (Fig. 2.4). At some wavelengths in the infrared the atmosphere – in the absence of clouds – is largely transparent, just as it is in the visible part of the spectrum. If our eyes were sensitive at these wavelengths we would be able to peer through the atmosphere to the sun, stars and moon above just as we can in the visible spectrum. At these wavelengths all the radiation originating from the Earth's surface leaves the atmosphere.

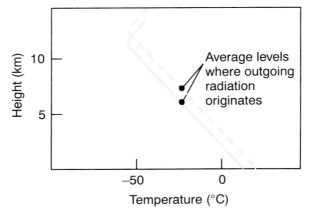

Fig. 2.3
The distribution of temperature in a convective atmosphere (full line). The dotted line shows how the temperature increases when the amount of carbon dioxide present in the atmosphere is increased (in the diagram the difference between the lines is exaggerated – for instance, for doubled carbon dioxide in the absence of other effects the increase in temperature is about 1.2 °C). Also shown for the two cases are the average levels from which thermal radiation leaving the atmosphere originates (about 6 km for the unperturbed atmosphere).

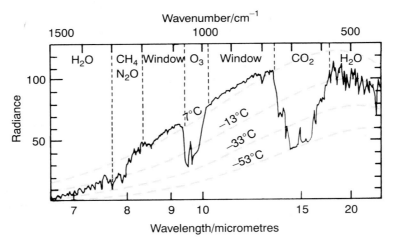

Fig. 2.4
Thermal radiation in the infrared (the visible part of the spectrum is between about 0.4 and 0.7 micrometres) emitted from the Earth's surface and atmosphere as observed over the Mediterranean Sea from a satellite instrument orbiting above the atmosphere, showing the parts of the spectrum where different gases contribute to the radiation[6]. Between wavelengths of about 8 and 14 micrometres, apart from the ozone band, the atmosphere, in the absence of clouds, is substantially transparent; this part of the spectrum is called a 'window' region. Superimposed on the spectrum are curves of radiation from a black body at 7 °C, −13 °C, −33 °C and −53 °C. The units of radiance are watts per square metre per steradian per wavenumber.

At other wavelengths radiation from the surface is strongly absorbed by some of the gases present in the atmosphere, in particular by water vapour and carbon dioxide.

Objects that are good absorbers of radiation are also good emitters of it. A black surface is both a good absorber and a good emitter, while a highly reflecting surface absorbs rather little and emits rather little too (which is why highly reflecting foil is used to cover the surface of a vacuum flask and why it is placed above the insulation in the lofts of houses).

Absorbing gases in the atmosphere absorb some of the radiation emitted by the Earth's surface and in turn emit radiation out to space. The amount of thermal radiation they emit is dependent on their temperature.

Radiation is emitted out to space by these gases from levels somewhere near the top of the atmosphere – typically from between 5 and 10 km high (see Fig. 2.3). Here, because of the convection processes mentioned earlier, the temperature is much colder – 30 to 50 °C or so colder – than at the surface. Because the gases are cold, they emit correspondingly less radiation. What these gases have done, therefore, is to absorb some of the radiation emitted by the Earth's surface but then to emit much less radiation out to space. They have, therefore, acted as a radiation blanket over the surface (note that the outer surface of a blanket is colder than inside the blanket) and helped to keep it warmer than it would otherwise be (Fig. 2.5).

There needs to be a balance between the radiation coming in and the radiation leaving the top of the atmosphere – as there was in the very simple model with which this chapter started. Figure 2.6 shows the various components of the radiation entering and leaving the top of the atmosphere for the real atmosphere situation. On average, 240 watts per square metre of solar radiation are absorbed by the atmosphere and the surface; this is less than the 288 watts mentioned at the beginning of the chapter, because now the effect of clouds is being taken into account. Clouds reflect some of the incident radiation from the sun back out to space. However, they also absorb and emit thermal radiation and have a blanketing effect similar to that of the greenhouse gases. These two effects work in opposite senses: one (the reflection of solar radiation) tends to cool the Earth's surface and the other (the absorption of thermal radiation) to warm it. Careful consideration of these two effects shows that on average the net effect of clouds on the total budget of radiation results in a slight cooling of the Earth's surface[7].

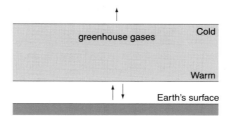

Fig. 2.5
The blanketing effect of greenhouse gases.

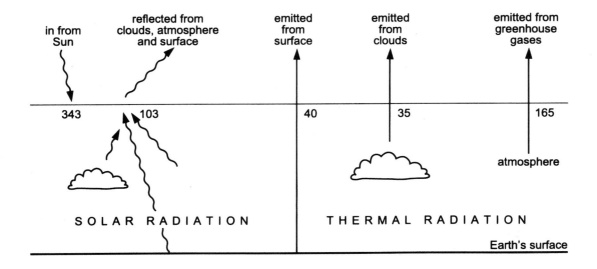

Fig. 2.6
Components of the radiation (in watts per square metre) which on average enter and leave the Earth's atmosphere and make up the radiation budget for the atmosphere.

The numbers in Fig. 2.6 demonstrate the required balance – 240 watts per square metre on average coming in and 240 watts per square metre on average going out. The temperature of the surface and hence of the atmosphere above adjusts itself to ensure that this balance is maintained. It is interesting to note that the greenhouse effect can only operate if there are colder temperatures in the higher atmosphere. Without the structure of decreasing temperature with height, therefore, there would be no greenhouse effect on the Earth.

Mars and Venus

Similar greenhouse effects also occur on our nearest planetary neighbours, Mars and Venus. Mars is smaller than the Earth and possesses, by Earth's standards, a very thin atmosphere. A barometer on the surface of Mars would record an atmospheric pressure less than 1 per cent of that on the Earth. Its atmosphere, which consists almost entirely of carbon dioxide, contributes a small but significant greenhouse effect.

The planet Venus, which can often be seen fairly close to the sun in the morning or evening sky, has a very different atmosphere to Mars. Venus is about the same size as the Earth. A barometer for use on Venus would need to survive very hostile conditions and would need to be able to measure a pressure about one hundred times as great as that on the Earth. Within the Venus atmosphere, which consists very largely of carbon dioxide, deep clouds consisting of droplets of almost pure sulphuric acid completely cover the planet and prevent most of the sunlight from reaching the surface. Some Russian space probes that have landed there have recorded what would be dusk-like conditions on the Earth – only 1 or 2 per cent of the sunlight present above the clouds penetrates that far. One might suppose, because of the small amount of solar energy available to keep the surface warm, that it would be rather cool. On the contrary; measurements from the same Russian space probes find a temperature there of about 525 °C – a dull red heat, in fact.

The reason for this very high temperature is the greenhouse effect. Because of the very thick absorbing atmosphere of carbon dioxide, very little of the thermal radiation from the surface can get out. The atmosphere acts as such an effective radiation blanket that, although there is not much solar energy to warm the surface, the greenhouse effect amounts to nearly 500 °C.

The 'runaway' greenhouse effect

What occurs on Venus is an example of what has been called the 'runaway' greenhouse effect. It can be explained by imagining the early history of the Venus atmosphere, which was formed by the release of gases from the interior of the planet. To start with it would contain a lot of water vapour, a powerful greenhouse gas (Fig. 2.7). The greenhouse effect of the water vapour would cause the temperature at the surface to rise. The increased temperature would lead to more evaporation of water from the surface, giving more atmospheric water vapour, a larger greenhouse effect and therefore a further increased surface temperature. The process would continue until either the atmosphere became saturated with water vapour or all the available water had evaporated.

A runaway sequence something like this seems to have occurred on Venus. Why, we may ask, has it not happened on the Earth, a planet of about the same size as Venus and, so far as is known, of a similar initial chemical composition? The reason is that Venus is closer to the sun than the Earth; the amount of solar energy per square metre falling on Venus is about twice that falling on the Earth. The surface of Venus, when there was no atmosphere, would have started off at a temperature of just over 50 °C (Fig. 2.7). Throughout the sequence described above for Venus, water on the surface would have been continuously boiling. Because of the high temperature, the atmosphere would never have become saturated with water vapour. The Earth, however, would have started at a colder temperature; at each stage of the sequence it would

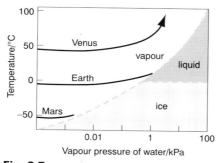

Fig. 2.7
Illustrating the evolution of the atmospheres of the Earth, Mars and Venus. In this diagram the surface temperatures of the three planets are plotted against the vapour pressure of water in their atmospheres as they evolved. Also on the diagram (dashed) are the phase lines for water, dividing the diagram into regions where vapour, liquid water or ice are in equilibrium. For Mars and the Earth the greenhouse effect is halted when water vapour is in equilibrium with ice or liquid water. For Venus no such halting occurs and the diagram illustrates the 'runaway' greenhouse effect[8].

have arrived at an equilibrium between the surface and an atmosphere saturated with water vapour. There is no possibility of such runaway greenhouse conditions occurring on the Earth.

The enhanced greenhouse effect

After our excursion to Mars and Venus, let us return to Earth! The natural greenhouse effect is due to the gases water vapour and carbon dioxide present in the atmosphere in their natural abundances as now on Earth. The amount of water vapour in our atmosphere depends mostly on the temperature of the surface of the oceans; most of it originates through evaporation from the ocean surface and is not influenced directly by human activity. Carbon dioxide is different. Its amount has changed substantially – by nearly 30 per cent so far – since the Industrial Revolution, due to human industry and also because of the removal of forests (see Chapter 3). Future projections are that, in the absence of controlling factors, the rate of increase in atmospheric carbon dioxide will accelerate and that its atmospheric concentration will double from its pre-industrial value well within the next hundred years (Fig. 3.5).

This increased amount of carbon dioxide is leading to global warming of the Earth's surface because of its enhanced greenhouse effect. Let us imagine, for instance, that the amount of carbon dioxide in the atmosphere suddenly doubled, everything else remaining the same (Fig. 2.8). What would happen to the numbers in the radiation budget I presented earlier (Fig. 2.6)? The solar radiation budget would not be affected. The greater amount of carbon dioxide in the atmosphere means that the thermal radiation emitted from it will originate on average from a higher and colder level than before (Fig. 2.3). The thermal radiation budget will therefore be reduced, the amount of reduction being about 4 watts per square metre.

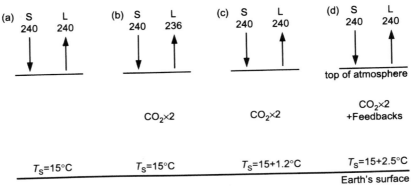

Fig. 2.8
Illustrating the enhanced greenhouse effect. Under natural conditions (a) the net solar radiation coming in (S = 240 watts per square metre) is balanced by thermal radiation (L) leaving the top of the atmosphere; average surface temperature (T_s) is 15 °C. If the carbon dioxide concentration is suddenly doubled (b), L is decreased by 4 watts per square metre. Balance is restored if nothing else changes (c) apart from the temperature of the surface and lower atmosphere which rises by 1.2 °C. If feedbacks are also taken into account (d) the average temperature of the surface rises by about 2.5 °C.

This causes a net imbalance in the overall budget of 4 watts per square metre. More energy is coming in than going out. To restore the balance the surface and lower atmosphere will warm up. If nothing changes apart from the temperature – in other words, the clouds, the water vapour, the ice and snow cover and so on, are all the same as before – the temperature change turns out to be about 1.2 °C.

In reality, of course, many of these other factors will change, some of them in ways that add to the warming (these are called positive feedbacks), others in ways that might reduce the warming (negative feedbacks). The situation is therefore much more complicated than this simple calculation. These complications will be considered in more detail in Chapter 5. Suffice it to say here that the best estimate at the present time of the increased average temperature of the Earth's surface if carbon dioxide levels were to be doubled is about twice that of the simple calculation: 2.5 °C. As the last chapter explained, for the global average temperature this is a large change. It is this global warming expected to result from the enhanced greenhouse effect which is the cause of current concern.

Having dealt with a doubling of the amount of carbon dioxide, it is interesting to ask what would happen if all the carbon dioxide were removed from the atmosphere. It is sometimes supposed that the outgoing radiation would be changed by 4 watts per square metre in the other direction and that the Earth would then cool by one or two degrees Celsius. In fact, that would happen if the carbon dioxide amount were to be halved. If it were to be removed altogether, the change in outgoing radiation would be around 25 watts per square metre – six times as big – and the temperature change would be similarly increased. The reason for this is that with the amount of carbon dioxide currently present in the atmosphere there is maximum carbon dioxide absorption over much of the region of the spectrum where it absorbs (Fig. 2.4), so that a big change in gas concentration leads to a relatively small change in the amount of radiation it absorbs[9]. This is like the situation in a pool of water: when it is clear, a small amount of mud will make it appear muddy, but when it is muddy, adding more mud will make little difference.

An obvious question to ask is: has evidence of the enhanced greenhouse effect been seen in the recent climatic record? Chapter 4 will look at the record of temperature on the Earth during the last century or so, during which the Earth has warmed on average by about half a degree Celsius. As we shall see in Chapters 4 and 5, although there are good reasons for supposing that some of this warming is due to the enhanced greenhouse effect, because of the size of natural climate variability it cannot yet be attributed unequivocally to that cause. Although we cannot yet be completely positive that we have seen it, we have good reasons for believing that the effect is real.

To summarize the argument so far:

- No one doubts the reality of the natural greenhouse effect, which keeps us over 20 °C warmer than we would otherwise be. The science of it is well understood; it is similar science which applies to the enhanced greenhouse effect.
- Substantial greenhouse effects occur on our nearest planetary neighbours, Mars and Venus. Given the conditions which exist on those planets, the

sizes of their greenhouse effects can be calculated, and good agreement has been found with those measurements which are available.

* The study of climates of the past gives some clues about the greenhouse effect, as Chapter 4 will show.

First, however, the greenhouse gases themselves must be considered. How does carbon dioxide get into the atmosphere, and what other gases affect global warming?

Questions

1   Carry out the calculation described in Note 4 (refer also to Note 2) which obtains an equilibrium average temperature of $-18\ ^\circ C$ for an Earth partially covered with clouds such that 30% of the incoming solar radiation is reflected. If clouds are assumed to cover half the Earth and if the reflectivity of the clouds increases by 1% what change will this make in the resulting equilibrium average temperature?

2   It is sometimes argued that the greenhouse effect of carbon dioxide is negligible because its absorption band in the infrared is so close to saturation that there is very little additional absorption of radiation emitted from the surface. What are the fallacies in this argument?

3   Use the information in Fig 2.4 to estimate approximately the surface temperature which would result if carbon dioxide were completely removed from the atmosphere. What is required is that the total energy radiated by the Earth plus atmosphere should remain the same, i.e. the area under the radiance curve in Fig 2.4 should be unaltered. On this basis construct a new curve with the carbon dioxide band absent[10].

4   Using information from books or articles on climatology or meteorology describe why the presence of water vapour in the atmosphere is of such importance in determining the atmosphere's circulation.

5   Estimates of regional warming due to increased greenhouse gases are generally larger over land areas than over ocean areas. What might be the reasons for this?

6   (for students with a background in physics) What is meant by Local Thermodynamic Equilibrium (LTE)[11], a basic assumption underlying calculations of radiative transfer in the lower atmosphere appropriate to discussions of the greenhouse effect? Under what conditions does LTE apply?

Notes

1   It is one quarter because the area of the Earth's surface is four times the area of the disc which is the projection of the Earth facing the sun – see Fig. 2.1.

2   The radiation by a black body is the Stefan-Boltzmann constant (5.67×10^{-8} J m^{-2} K^{-4} s^{-1}) multiplied by the fourth power of the body's absolute temperature in degrees Kelvin. The absolute temperature is the temperature in degrees Celsius plus 273 (one degree K = one degree C).

3   These calculations using a simple model of an atmosphere containing nitrogen and oxygen only have been carried out to illustrate the effect of the other gases, especially water vapour and carbon dioxide. It is not, of course, a model that can exist in

reality. All the water vapour could not be removed from the atmosphere above a water or ice surface. Further, with an average surface temperature of –6 °C, in a real situation the surface would have much more ice cover. The additional ice would reflect more solar energy out to space leading to a further lowering of the surface temperature.

4 The above calculation is often carried out using a figure of 30 per cent for the average reflectivity of the Earth and atmosphere rather than the 16 per cent assumed here; the calculation of surface temperature then gives –18 °C for the average surface temperature rather than the –6 °C found here. The higher figure of 30 per cent for the Earth's average reflectivity is applicable when clouds are also included, in which case the average temperature of –18 °C is not applicable to the Earth's surface but to some appropriate level in the atmosphere. Further, clouds not only reflect solar radiation but also absorb thermal radiation, and so have a blanketing effect similar to greenhouse gases. For the purpose of illustrating the effect of greenhouse gases, therefore, it is more correct to omit the effect of clouds from this initial calculation.

5 Further details can be found in F. B. Mudge, 'The development of the greenhouse theory of global climate change from Victorian times', *Weather*, **52**, 1997, pp. 13–16.

6 Spectrum taken with the infrared interferometer spectrometer flown on the satellite Nimbus 4 in 1971 and described by R. A. Hanel *et al., Appl. Opt.,* **10**, 1971, pp. 1376–82.

7 More detail of the radiative effects of clouds is given in Chapter 5 – see Figs. 5.14 and 5.15.

8 From J. T. Houghton, *The Physics of Atmospheres*, 2nd edn, CUP, 1986.

9 The dependence of the absorption on the concentration of the gas is approximately logarithmic.

10 For some helpful diagrams and more information about the infrared spectrum of different greenhouse gases, see J. E. Harries, 'The greenhouse Earth: a view from space' *Quart. J. R. Met. Soc.*, **122**, 1996, pp. 799–818.

11 For information about LTE see, for instance, J. T. Houghton, *The Physics of Atmospheres*, 2nd edn, CUP, 1986.

3 The Greenhouse Gases

The greenhouse gases are those gases in the atmosphere which, by absorbing thermal radiation emitted by the Earth's surface, have a blanketing effect upon it. The most important of the greenhouse gases is water vapour, but its amount in the atmosphere is not changing directly because of human activities. The important greenhouse gases which are directly influenced by human activities are carbon dioxide, methane, nitrous oxide, the chlorofluorocarbons (CFCs) and ozone. This chapter will describe what is known about the origin of these gases, how their concentration in the atmosphere is changing and how it is controlled. Also considered will be particles in the atmosphere of anthropogenic origin which can act to cool the surface.

Which are the most important greenhouse gases?

Figure 2.4 illustrated the regions of the infrared spectrum where the greenhouse gases absorb. Their importance as greenhouse gases depends both on their concentration in the atmosphere (Table 2.1) and on the strength of their absorption of infrared radiation. Both these quantities differ greatly for various gases.

Carbon dioxide is the most important of the greenhouse gases which are increasing in atmospheric concentration because of human activities. If, for the moment, we ignore the effects of the CFCs and of changes in ozone, which vary considerably over the globe and which are difficult to quantify, the increase in carbon dioxide (CO_2) has contributed about 70 per cent of the enhanced greenhouse effect to date, methane (CH_4) about 24 per cent, and nitrous oxide (N_2O) about 6 per cent (Fig. 3.8).

Radiative forcing

In this chapter we shall use the concept of *radiative forcing* to compare the relative greenhouse effects of different atmospheric constituents. It is necessary therefore at the start to define radiative forcing.

In Chapter 2 we noted that, if the carbon dioxide in the atmosphere were suddenly doubled, everything else remaining the same, a net radiation imbalance near the top of the atmosphere of 4 Wm^{-2} would result. This radiation imbalance is an example of radiation forcing which is defined as the change in average net radiation at the top of the troposphere[1] (the lower atmosphere; for definition see Glossary) which occurs because of a change in the concentration of a greenhouse gas or because of some other change in the overall climate system; for instance a change in the incoming solar radiation would constitute a radiative forcing. As we saw in the discussion in Chapter 2, over time the climate responds to restore the radiative balance between incoming and outgoing radiation. A positive radiative forcing tends on average to warm the surface and a negative radiative forcing tends on average to cool the surface.

Carbon dioxide and the carbon cycle

Carbon dioxide provides the dominant means through which carbon is transferred in nature between a number of natural carbon reservoirs – a process known as the carbon cycle. We contribute to this cycle every time we breathe. Using the oxygen we take in from the atmosphere, carbon from our food is burnt and turned into carbon dioxide which we then exhale; in this way we are provided with the energy we need to maintain our life. Animals contribute to atmospheric carbon dioxide in the same way; so do fires, rotting wood and decomposition of organic material in the soil and elsewhere. To offset these processes of respiration whereby carbon is turned into carbon dioxide, there are processes involving photosynthesis in plants and trees which work the opposite way; in the presence of light, they take in carbon dioxide, use the carbon for growth and let out the oxygen back into the atmosphere. Both respiration and photosynthesis also occur in the ocean.

Figure 3.1 is a simple diagram of the way carbon cycles between the various reservoirs – the atmosphere, the oceans (including the ocean biota), the soil and the land biota (biota is a word that covers all living things – plants, trees,

Fig. 3.1
The reservoirs of carbon in the Earth, the biosphere, the ocean and the atmosphere and the annual exchanges of carbon dioxide (expressed in terms of mass of carbon it contains) between the reservoirs[2]. The units are thousand millions of tonnes or gigatonnes (Gt).

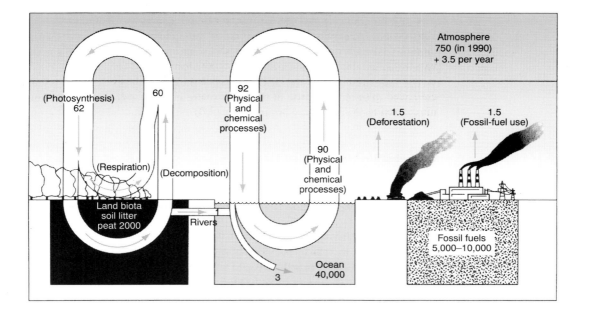

animals and so on – on land and in the ocean, which make up a whole known as the biosphere). The diagram shows that the movements of carbon (in the form of carbon dioxide) into and out of the atmosphere are quite large; about one quarter of the total amount in the atmosphere is cycled in and out each year, half of this with the land biota and the other half through physical and chemical processes across the ocean surface. The land and ocean reservoirs are much larger than the amount in the atmosphere; small changes in these larger reservoirs could therefore have a large effect on the atmospheric concentration; the release of just 2 per cent of the carbon stored in the oceans would double the amount of atmospheric carbon dioxide.

It is important to realise that on the timescales with which we are concerned anthropogenic carbon emitted into the atmosphere is not destroyed but redistributed among the various carbon reservoirs. Carbon dioxide is therefore different from other greenhouse gases which are destroyed by chemical action in the atmosphere. The carbon reservoirs exchange carbon between themselves on a wide range of timescales determined by their respective turnover times – which range from less than a year to decades (for exchange with the top layers of the ocean and the land biosphere) to millennia (in the deep ocean or long-lived soil pools). These timescales are generally much longer than the average time a particular carbon dioxide molecule spends in the atmosphere, which is only about 4 years. The large range of turnover times means that the time taken for a perturbation in the atmospheric carbon dioxide concentration to relax back to an equilibrium cannot be described by a single time constant. Although a lifetime of about a hundred years is often quoted for atmospheric carbon dioxide so as to provide some guide, use of a single lifetime can be very misleading.

Before human activities became a significant disturbance, and over periods short compared with geological timescales, the exchanges between the reservoirs were remarkably constant. For several thousand years before the beginning of industrialization around 1750, a steady balance was maintained, such that the concentration of carbon dioxide in the atmosphere as measured from ice cores (see Chapter 4) kept within about ten parts per million of a mean value of about 280 parts per million by volume (ppmv).

The Industrial Revolution disturbed this balance and has resulted in a concentration of carbon dioxide in the atmosphere which has increased by about 30 per cent, from 280 ppmv around the year 1700 to a value of over 360 ppmv at the present day (Fig. 3.2). Accurate measurements which have been made since 1959 from an observatory near the summit of Mauna Loa in Hawaii show that carbon dioxide is currently increasing on average each year by about 1.5 ppmv (Fig. 3.2b), although there are large variations from year to year (Fig 3.2c). This increase spread through the atmosphere adds about 3.3 thousand million tonnes (or gigatonnes, Gt) to the atmospheric carbon reservoir each year.

It is easy to establish how much coal, oil and gas are being burnt worldwide each year. Most of it is to provide energy for human needs: for heating and domestic appliances, for industry and for transport (considered in detail in Chapter 11). The amount of these fossil fuels burnt has increased rapidly since the Industrial Revolution (Fig. 3.3); currently the total contains about 6 Gt of

Fig. 3.2
(a) The increase of atmospheric carbon dioxide since 1700[3] showing measurements from ice cores in Antarctica (squares) and since 1957, direct measurements from the Mauna Loa observatory in Hawaii (triangles). (b) More detailed measurements of the increase of carbon dioxide since 1959 as observed at Mauna Loa (showing the annual cycle) and at the South Pole[4]. (c) Growth rate of carbon dioxide concentration since 1958 at Mauna Loa[5]. The smooth curve shows the same data but filtered to suppress variations on timescales less than approximately 10 years. Note the small rate of growth from 1991 to 1994 the reasons for which are not well understood but it may be connected with events such as the Pinatubo volcanic eruption in 1991 or the very unusual El Niño since 1990.

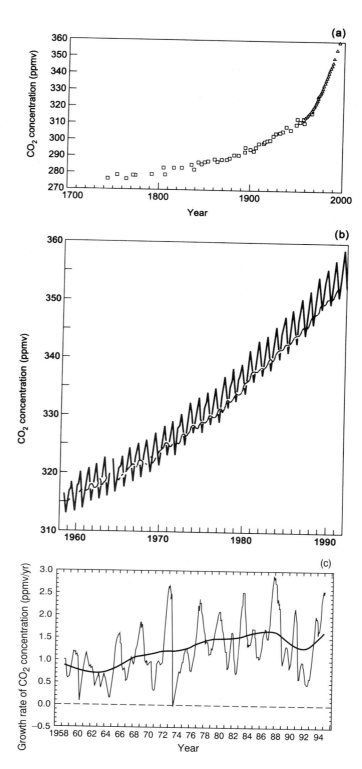

Table 3.1
Components of annual average anthropogenic carbon budget for 1980 to 1989 – in Gt of carbon per year[6].

CO_2 sources	
(1) Emissions from fossil fuel combustion and cement production	5.5 ± 0.5*
(2) Net emissions from changes in tropical land-use	1.6 ± 1.0[@]
(3) Total anthropogenic emissions = (1) + (2)	7.1 ± 1.1

Partitioning amongst reservoirs	
(4) Storage in the atmosphere	3.3 ± 0.2
(5) Ocean uptake	2.0 ± 0.8
(6) Uptake by northern hemisphere forest regrowth	0.5 ± 0.5[#]
(7) Inferred sink: 3−(4+5+6)	1.3 ± 1.5[§]

Notes:
* For comparison, emissions in 1994 were 6.1 GtC/yr.
[@] Consistent with Chapter 24 of IPCC WGII (1995).
[#] This number is consistent with the independent estimate, given in IPCC WGII (1995), of 0.7 ± 0.2 GtC/yr for the mid- and high latitude forest sink.
[§] This inferred sink is consistent with independent estimates, given in Chapter 9 of IPCC WGI (1995), of carbon uptake due to nitrogen fertilisation (0.5 ± 1.0 GtC/yr), plus the range of other uptakes 0−2 GtC/yr due to CO_2 fertilisation and climatic effects.

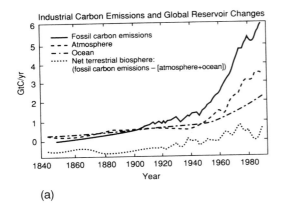

(a)

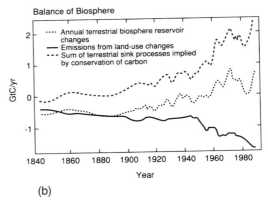

(b)

Fig. 3.3
(a) Fossil carbon emissions (based on statistics of fossil fuel and cement production) and representative calculations of global reservoir changes: atmosphere (deduced from direct observations and ice core measurements), ocean (calculated with the GFDL ocean carbon model) and net terrestrial biosphere (calculated as remaining imbalance). The calculation implies that the terrestrial biosphere was a net source to the atmosphere prior to 1940 (negative values) and a net sink since about 1960. (b) Contributions to the carbon balance of the terrestrial biosphere[7]. Emissions from land use changes (including tropical deforestation) are plotted negative because they represent a loss of biospheric carbon.

carbon, nearly all of which enters the atmosphere as carbon dioxide. It is not so easy to estimate the amount of carbon dioxide released to the atmosphere because of changes of land use, in particular the burning and decay of forests. The range of estimates averaged over the decade of the 1980s is given in Table 3.1; other data is presented in Fig 3.3. If we assume a middle value of 1.5 Gt per year from changes in land use and deforestation and add to it the 6 Gt per year from fossil fuel burning, the total entering the atmosphere from human activities sums to about 7.5 Gt per year. Since the annual net increase in the atmosphere is about 3.3 Gt, about 45 per cent of the 7.5 Gt of new carbon remains to increase the atmospheric concentration. The other 55 per cent is taken up between the other two reservoirs: the oceans and the land biota.

About 95 per cent of fossil fuel burning occurs in the northern hemisphere, so there is more carbon dioxide there than in the southern hemisphere. The difference is currently about 2 parts per million (Fig. 3.2b) and, over the years, has grown in parallel with fossil fuel emissions, thus adding further compelling evidence that the atmospheric increase in carbon dioxide levels results from these emissions.

We turn now to what happens in the oceans. We know that carbon dioxide dissolves in water: carbonated drinks make use of that fact. Carbon dioxide is continually being exchanged with the air above the ocean across the whole ocean surface (about 90 Gt per year is so exchanged – Fig 3.1), particularly as waves break. An equilibrium is set up between the concentration of carbon dioxide dissolved in the surface waters and the concentration in the air above the surface. The chemical laws governing this equilibrium are such that if the atmospheric concentration changes by 10 per cent the concentration in solution in the water changes by only one tenth of this: 1 per cent.

This change will occur quite rapidly in the upper waters of the ocean, the top hundred metres or so, so enabling part of the anthropogenic (i.e. human-generated) carbon dioxide added to the atmosphere (most of the ocean's share of the 55 per cent mentioned above) to be taken up quite rapidly. Absorption in the lower levels in the ocean takes longer; mixing of surface water with water at lower levels takes up to several hundred years or for the deep ocean over a thousand years. So the oceans do not provide as immediate a sink for increased atmospheric carbon dioxide as might be suggested by the size of the exchanges with the large ocean reservoir. For short-term changes only the surface layers of water play a large part in the carbon cycle.

Biological activity in the oceans also plays an important role. It may not be immediately apparent, but the oceans are literally teeming with life. Although the total mass of living matter within the oceans is not large, it has a high rate of turnover. Living material in the oceans is produced at some 30–40 per cent of the rate of production on land. Most of this production is of plant and animal plankton which go through a rapid series of life cycles. As they die and decay some of the carbon they contain is carried downwards into lower levels of the ocean adding to the carbon content of those levels. Some is carried to the very deep water or to the ocean bottom where, so far as the carbon cycle is concerned, it is out of circulation for hundreds or thousands of years. This process, whose contribution to the carbon cycle is known as the 'biological pump' (see box), has been important in determining the changes of carbon dioxide concentration in both the atmosphere and the ocean during the ice ages (see Chapter 4).

Computer models – which calculate solutions for the mathematical equations describing a given physical situation, in order to predict its behaviour – have been set up to describe in detail the exchanges of carbon between the

---

The 'biological pump' in the oceans[8]

In temperate and high latitudes there is a peak each spring in ocean biological activity. During the winter, water rich in nutrients is transferred from deep water to levels near the surface. As sunlight increases in the spring an explosive growth of the plankton population occurs, known as the 'spring bloom'. Pictures of the colour of the ocean taken from satellites orbiting the Earth can demonstrate dramatically where this is happening.

Plankton are small plants (phytoplankton) and animals (zooplankton) which live in the surface waters of the ocean; they range in size between about one thousandth of a millimetre across and the size of typical insects on land. Herbivorous zooplankton graze on phytoplankton; carnivorous zooplankton eat herbivorous zooplankton. Plant and animal debris from these living systems sinks in the ocean. While sinking some decomposes and returns to the water as nutrients, some (perhaps about 1 per cent) reaches the deep ocean or the ocean floor, where it is lost to the carbon cycle for hundreds, thousands or even millions of years. The net effect of the 'biological pump' is to move carbon from the surface waters to lower levels in the ocean. As the amount of carbon in the surface waters is reduced, more carbon dioxide from the atmosphere can be drawn down in order to restore the surface equilibrium. It is thought that the 'biological pump' has remained substantially constant in its operation during the last century unaffected by the increase in carbon dioxide availability.

Evidence of the importance of the 'biological pump' comes from the paleoclimate record from ice cores (see Chapter 4). One of the constituents from the atmosphere trapped in bubbles in the ice is the gas methyl sulphonic acid which originates from decaying ocean plankton; its concentration is therefore an indicator of plankton activity. As the global temperature

began to increase when the last ice age receded nearly 20,000 years ago and as the carbon dioxide in the atmosphere began to increase (Fig. 3.2), the methyl sulphonic acid concentration decreased. An interesting link is thereby provided between the carbon dioxide in the atmosphere and marine biological activity. During the cold periods of the ice ages, enhanced biological activity in the ocean could have been responsible for maintaining the atmospheric carbon dioxide at a lower level of concentration – the 'biological pump' was having an effect.

There is some evidence from the paleo record of the biological activity in the ocean being stimulated by the presence of dust containing iron blown over the oceans from the land surface. This has led to some proposals in recent years to enhance the 'biological pump' through artificially introducing iron over suitable parts of the ocean. While an interesting idea, it seems from careful studies that even a very large scale operation would not have large practical effect.

The question then remains as to why the ice ages should be periods of greater marine biological activity than the warm periods in between. A British oceanographer, Professor John Woods, has suggested that the key may lie in what happens in the winter as nutrients are fed into the upper ocean ready for the spring bloom. When there is less atmospheric carbon dioxide, the cooling by radiation from the surface of the ocean increases. Since convection in the upper layers of the ocean is driven by cooling at the surface, the increased cooling results in a greater depth of the mixed layer near the top of the ocean where all the biological activity occurs. This is an example of a positive biological feedback; a greater depth of layer means more plankton growth. Woods calls it the 'plankton multiplier'[9].

atmosphere and different parts of the ocean. To test the validity of these models, they have also been applied to the dispersal in the ocean of the carbon isotope ^{14}C which entered the ocean after the nuclear tests of the 1950s; the models simulate this dispersal quite well. From the model results, it is estimated that about 2 Gt (± 0.8 Gt) of the carbon dioxide added to the atmosphere each year ends up in the oceans (Table 3.1 and Fig 3.3). Observations of the relative distribution of the other isotopes of carbon in the atmosphere and in the oceans also confirm this estimate (see box).

To complete the carbon cycle budget, there needs to be added the effects of changes in the land biosphere other than those due to tropical deforestation which have already been mentioned. One of these is the uptake of carbon due to the regrowth of northern hemisphere forests which has been estimated at about 0.5 Gt per year (Table 3.1). Other processes which can increase the uptake of carbon in the land biosphere are the carbon dioxide 'fertilization' effect (increased carbon dioxide in the atmosphere leads to increased growth in

---

What we can learn from carbon isotopes

Isotopes are chemically identical forms of the same element but with different atomic weights. Three isotopes of carbon are important in studies of the carbon cycle: the most abundant isotope ^{12}C which makes up 98.9 per cent of ordinary carbon, ^{13}C present at about 1.1 per cent and the radioactive isotope ^{14}C which is present only in very small quantities. About ten kilograms of ^{14}C are produced in the atmosphere each year by the action of particle radiation from the sun; half of this will decay into nitrogen over a period of 5,730 years (the 'half-life' of ^{14}C).

When carbon in carbon dioxide is taken up by plants and other living things, less ^{13}C is taken up in proportion than ^{12}C. Fossil fuel such as coal and oil was originally living matter so also contains less ^{13}C (by about 18 parts per thousand) than the carbon dioxide in ordinary air in the atmosphere today. Adding carbon to the atmosphere from burning forests, decaying vegetation or fossil fuel will therefore tend to reduce the proportion of ^{13}C.

Because fossil fuel has been stored in the Earth for much longer than 5,730 years (the half-life of ^{14}C) it contains no ^{14}C at all. As carbon from fossil fuel is added to the atmosphere, therefore, the proportion of ^{14}C in the atmosphere is also reduced.

By studying the ratio of the different isotopes of carbon in the atmosphere, in the oceans, in gas trapped in ice cores and in tree rings, it is possible to find out where the additional carbon dioxide in the atmosphere has come from and also what amount has been transferred to the ocean. For instance, it has been possible to estimate for different times how much carbon dioxide has entered the atmosphere from the burning or decay of forests and other vegetation and how much from fossil fuels.

Similar isotopic measurements on the carbon in atmospheric methane provide information about how much methane from fossil fuel sources has entered the atmosphere at different times.

some plants – see box in Chapter 7), the effects of the use of nitrogen fertilizers and the effects of some changes in climate. The magnitude of these is difficult to estimate directly but their total of about 1.5 Gt per year, with a large range of uncertainty (Table 3.1 and Fig 3.3), can be inferred from the requirement to balance the overall carbon cycle budget.

A clue to the uptake of carbon by the land biosphere is provided from observations of the atmospheric concentration of carbon dioxide which, each year, show a regular cycle (Fig. 3.2b). Carbon dioxide is removed from the atmosphere during the growing season and returned as the vegetation dies away in the winter. Since there is more vegetation growth in the northern hemisphere than the southern, a minimum in the annual cycle of carbon dioxide in the atmosphere occurs in the northern summer. Estimates from carbon cycle models of the uptake by the land biosphere are constrained by these observations of the difference between the hemispheres.

Recent confirmation of the broad partitioning of added atmospheric carbon dioxide between the atmosphere, the oceans and the land biota as presented in Table 3.1, comes from trends in the atmospheric oxygen/nitrogen ratio as measured at various latitudes. These data are consistent with a budget for the period 1991–94 in which the global oceans and the northern land biota each removed the equivalent of approximately 30 per cent of fossil fuel emissions[10].

The carbon dioxide fertilization effect is an example of a biological feedback process. It is a negative feedback because, as carbon dioxide increases, it tends to increase the take-up of carbon dioxide by plants and therefore to reduce the amount in the atmosphere, decreasing the rate of global warming. Positive feedback processes, which would tend to accelerate the rate of global warming, also exist; in fact there are more potentially positive processes than negative ones (see box). Although scientific knowledge cannot yet put precise figures on them, some of the positive feedbacks could be large, especially if carbon dioxide were to continue to increase, with its associated global warming, through the twenty-first century into the twenty-second.

We now consider in more detail the effect on the concentration of atmospheric carbon dioxide of future anthropogenic emissions. In these considerations, the long time constants associated with the response of atmospheric carbon dioxide to change have important implications. Suppose, for instance, that all emissions into the atmosphere from human activities were suddenly halted. No sudden change would occur in the atmospheric concentration, which would decline only slowly. We could not expect it to approach its pre-industrial value for several hundred years.

But emissions of carbon dioxide are not halting, nor are they slowing; their increase is, in fact, becoming larger each year. Later chapters will present estimates of climate change next century due to the increase in greenhouse gases. A prerequisite for such estimates is the knowledge of what changes in carbon dioxide emissions there are likely to be. Estimating what will happen in the future is, of course, not easy. Because nearly everything we do has an influence on the emissions of carbon dioxide, it means estimating how human beings will behave and what their activities are likely to be. For instance, assumptions need to be made about population growth, economic growth, energy use, the development of energy sources and the likely influence of pressures to preserve

---

Feedbacks in the biosphere

As the greenhouse gases carbon dioxide and methane are added to the atmosphere because of human activities, biological or other feedback processes occurring in the biosphere influence the rate of increase of the atmospheric concentration of these gases. These processes will either tend to add to the anthropogenic increase (positive feedbacks) or to subtract from it (negative feedbacks).

Two feedbacks, one positive (the plankton multiplier in the ocean) and one negative (carbon dioxide fertilization) have already been mentioned in the text. Three other positive feedbacks are potentially important, although our knowledge is currently insufficient to quantify them at all precisely.

One is the effect of higher temperatures on respiration, especially of microbes in soils, leading to increased carbon dioxide emissions. A second is the reduction of growth or the die-back especially in forests because of the stress caused by climate change (see Chapter 7). The third positive feedback is the release of methane, as temperatures increase – from wetlands and from very large reservoirs of methane trapped in sediments in a hydrate form (tied to water molecules when under pressure) – mostly at high latitudes. Methane has been generated from the decomposition of organic matter present in these sediments over many millions of years. Because of the depth of the sediments this latter feedback is unlikely to become operative to a significant extent during the next century. However, were global warming to continue to increase substantially for more than a hundred years, it is estimated that releases from hydrates could become the largest single contributor of methane emissions into the atmosphere.

---

the environment. These assumptions are required for all countries of the world, both developing as well as developed ones. Further, since any assumptions made are unlikely to be fulfilled accurately in practice, it is necessary to make a variety of different assumptions, so that we can get some idea of the range of possibilities. Such possible futures are called *scenarios*.

The carbon dioxide emissions which would result from different scenarios prepared by the Intergovernmental Panel on Climate Change (IPCC)[11] and the World Energy Council (WEC)[12] are shown in Figure 3.4. The IPCC scenario IS 92a assumes a world population growing from 5.2 billion in 1990 to more than double that number, 11.3 billion, in 2100[13], moderate economic growth and no strong pressure to reduce carbon dioxide emissions for environmental reasons. In the absence of strong environmental action it can perhaps be considered as the most likely scenario – the 'business-as-usual' scenario. It shows close to a threefold rise in total emissions during next century (Table 3.2), made up of a large increase from the energy sector and a decrease of emissions from deforestation as less forests remain to be diminished. Details of the assumptions regarding the emissions from energy production for the WEC scenarios are given in Chapter 11 (see Table 11.2).

Table 3.2
Assumptions regarding annual emissions from different gases for IPCC scenario IS 92a.
The emissions labelled 'energy' include those from cement production (about 0.2 Gt in
1990). The gases SO_x are precursors to the formation of sulphate aerosols.

	1990	2025	2100
CO_2 (Gt carbon)			
from energy from fossil fuels	6.2	11.1	20.4
from deforestation	1.3	1.1	−0.1
Total	7.5	12.2	20.3
CH_4 (Tg)	506	659	917
N_2O (Tg nitrogen)	12.9	15.8	17.0
SO_x (Tg sulphur)	98	141	169

To turn the emission scenarios into future projections of atmospheric carbon
dioxide concentrations, they need to be incorporated into a computer model
(see also Chapters 5 and 6) of the carbon cycle which includes descriptions of

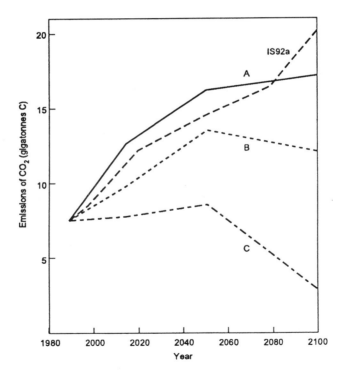

Fig. 3.4
Scenarios of carbon dioxide emissions next century from energy
generation from fossil fuels for IPCC scenario IS 92a and WEC
scenarios A, B and C (details in Table 11.2). The emissions from
WEC scenario B1 are not plotted separately; up to the year 2020
they are almost identical with those from IS 92a.

Fig. 3.5
Concentrations of carbon dioxide in the atmosphere resulting from the emission scenarios of Fig. 3.4[14] together with emissions from deforestation (Table 3.2). Also shown is the concentration curve for a scenario in which the annual emissions of carbon dioxide from fossil fuel emissions are kept constant beyond the year 2000 at 7.1Gt of carbon. The shaded area around the curve for the IS 92a scenario provides an estimate of the uncertainty in the model computations. Similar uncertainty applies to the other scenarios.

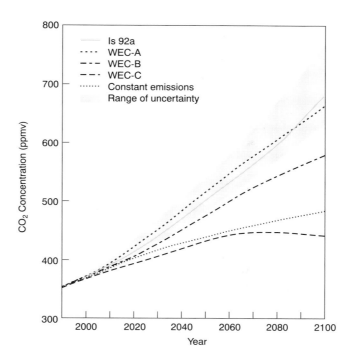

all the exchanges already mentioned. A number of scientists in different countries have developed carbon cycle models. Figure 3.5 shows the results from the application of one such model to a number of different emission scenarios[15]. Other models give similar results. Figure 3.5 also provides an indication of the uncertainty in the model arising from inadequate understanding of the carbon cycle and of the details of the carbon cycle budget.

The information in Figs. 3.4 and 3.5 provides essential input for our discussion in later chapters of the consequences of increasing greenhouse gas emissions. Chapter 6 looks at the likely climate change for the IS 92a 'business-as-usual' scenario; later on Chapter 10 considers the changes which may occur under some of the other scenarios. The WEC scenario C is particularly interesting as, of the scenarios we have mentioned, it is the only one in which the increase in carbon dioxide concentration in the atmosphere is halted before the end of the next century and a stable level reached.

Other greenhouse gases

Methane

Methane is the main component of natural gas. Its common name used to be marsh gas because it can be seen bubbling up from marshy areas where organic material is decomposing. Data from ice cores show that for at least two thousand years before 1800 its concentration in the atmosphere was about 0.7 ppmv. Since then its concentration has more than doubled and is increasing on average at about 0.6 per cent per year (Fig. 3.6)[16]. Although the concentration of methane in the atmosphere is much less than that of carbon dioxide (less than 2 ppmv compared with about 350 ppmv for carbon dioxide), its greenhouse effect is far from negligible. That is because the enhanced greenhouse effect caused by a molecule of methane is about 7.5 times that of a molecule of carbon dioxide[17].

The main natural source of methane is from wetlands. A variety of other sources result directly or indirectly from human activities, for instance from leakage from natural gas pipelines and from oil wells, from generation in rice paddy fields, from enteric fermentation (belching) from cattle and other livestock, from the decay of rubbish in landfill sites and from wood and peat burning. Details of the best estimates of the sizes of these sources are shown in Table 3.3. Attached to many of the numbers is a wide range of uncertainty. It is, for instance, difficult to estimate the amount produced in paddy fields averaged on a worldwide basis. The amount varies enormously during the rice growing season and also very widely from region to region. Similar problems arise when trying to estimate the amount produced by animals. Measurements of the proportions of the different isotopes of carbon (see box on p 29) in atmospheric methane assist considerably in helping to tie down the proportion which comes from fossil fuel sources, such as leakage from mines and from natural gas pipelines.

The main process for the removal of methane from the atmosphere is through chemical destruction. It reacts with hydroxyl (OH) radicals, which are present in the atmosphere because of processes involving sunlight, oxygen, ozone and water vapour. The average lifetime of methane in the atmosphere is determined by the rate of this loss process. At about 12 years[18], it is much shorter than the lifetime of carbon dioxide.

Although methane sources cannot be identified very precisely, the largest sources apart from natural wetlands are closely associated with human activities. It is interesting to note that the increase of atmospheric methane (Fig. 3.6)

Fig. 3.6
Concentration of methane in the atmosphere over the past few centuries as measured from air trapped in ice cores[19]. The values (full circles) for 1992 and 1995 are also plotted.

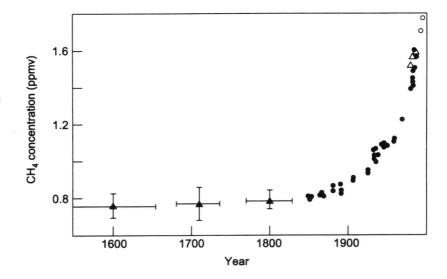

Table 3.3
Estimated sources and sinks of methane in millions of tonnes per year[20]. The first column shows the best estimate from each source; the second column illustrates the uncertainty in the estimates by giving a range of values.

Sources		
NATURAL:		
Wetlands	115	(55–150)
Termites	20	(10–50)
Ocean	10	(5–50)
Other	15	(10–40)
HUMAN-GENERATED:		
Coal mining, natural gas, petroleum industry	100	(70–120)
Rice paddies	60	(20–100)
Enteric fermentation	85	(65–100)
Animal wastes	25	(20–70)
Domestic sewage treatment	25	(15–80)
Landfills	40	(20–70)
Biomass burning	40	(20–80)
Sinks		
Atmospheric removal	530	(440–625)
Removal by soils	30	(15–45)
Atmospheric increase	37	(35–40)

follows very closely the growth of human population since the Industrial Revolution. If no attempt is made to control human-related sources of methane it is reasonable to assume that this trend will continue. On the basis of United Nations estimates that the world's population will approximately double next century, we can estimate that, unless efforts are made to reduce human-related methane sources, the enhanced greenhouse effect of increased methane (which to date is about 20 per cent of that of increased carbon dioxide through human activity) will also approximately double over that time (Fig. 3.9).

Nitrous oxide
Nitrous oxide, used as a common anaesthetic and known as laughing gas, is another minor greenhouse gas. Its concentration in the atmosphere of about 0.3 ppmv is rising at about 0.25 per cent per year and is about 13 per cent greater than in pre-industrial times. It possesses a relatively long atmospheric lifetime of about 120 years. The additional sources which are leading to its increase are not well identified although the chemical industry (for example, nylon production), deforestation and agricultural practices all play some part. To estimate its likely increase next century we need first to know much more detail about the human-generated sources responsible for its production.

Chlorofluorocarbons (CFCs) and ozone
The CFCs are man-made chemicals which, because they vaporize just below room temperature and because they are non-toxic and non-flammable, appear

to be ideal for use in refrigerators, the manufacture of insulation and aerosol spray cans. Since they are so chemically unreactive, once they are released into the atmosphere they remain for a long time – one or two hundred years – before being destroyed. As their use has increased rapidly through the 1980s their concentration in the atmosphere has been building up so that they are now present (adding together all the different CFCs) in about 1ppbv (part per thousand million – or billion – by volume). This may not sound very much, but it is quite enough to cause two serious environmental problems.

The first problem is that they destroy ozone. Ozone (O_3), a molecule consisting of three atoms of oxygen, is an extremely reactive gas present in small quantities in the stratosphere (a region of the atmosphere between about 10 km and 50 km in altitude). Ozone molecules are formed through the action of ultraviolet radiation from the sun on molecules of oxygen. They are in turn destroyed by a natural process as they absorb solar ultraviolet radiation at slightly longer wavelengths – radiation which would otherwise be harmful to us and to other forms of life at the Earth's surface. The amount of ozone in the stratosphere is determined by the balance between these two processes, one forming ozone and one destroying it. What happens when CFC molecules move into the stratosphere is that some of the chlorine atoms they contain are stripped off, also by the action of ultraviolet sunlight. These chlorine atoms readily react with ozone, reducing it back to oxygen and adding to the rate of destruction of ozone. This occurs in a catalytic cycle – one chlorine atom can destroy many molecules of ozone.

The ozone hole
The problem of ozone destruction was brought to world attention in 1985 when Joe Farman, Brian Gardiner and Jonathan Shanklin at the British Antarctic Survey discovered a region of the atmosphere over Antarctica where, during the southern spring, about half the ozone overhead disappeared. The existence of the 'ozone hole' was a great surprise to the scientists; it set off an intensive investigation into its causes. The chemistry and dynamics of its formation turned out to be complex. They have now been unravelled, at least as far as their main features are concerned, leaving no doubt that chlorine atoms introduced into the atmosphere by human activities are largely responsible. Not only is there depletion of ozone in the spring over Antarctica (and to a lesser extent over the Arctic) but also substantial reduction, of the order of 5 per cent, of the total column of ozone – the amount above one square metre at a given point on the Earth's surface – at mid-latitudes in both hemispheres.

Because of these serious consequences of the use of CFCs, international action has been taken. Many governments have now signed the Montreal Protocol set up in 1987 which, together with the Amendments agreed in London in 1991 and in Copenhagen in 1992, requires that manufacture of CFCs be phased out completely by the year 1996 in industrialized countries and by 2006 in developing countries. Because of this action the concentration of CFCs in the atmosphere is no longer increasing. However, since they possess a long life in the atmosphere, little decrease will be seen for some time and substantial quantities will be present well over a hundred years from now.

So much for the problem of ozone destruction. The other problem with CFCs and ozone, the one which concerns us here, is that they are both greenhouse gases. They possess absorption bands in the region known as the long-wave atmospheric window (see Fig. 2.4) where few other gases absorb. Because, as we have seen, the CFCs destroy some ozone, the greenhouse effect of the CFCs is partially compensated by the reduced greenhouse effect of atmospheric ozone.

First considering the CFCs on their own, a CFC molecule added to the atmosphere has a greenhouse effect five to ten thousand times greater than an added molecule of carbon dioxide. Thus, despite their very small concentration compared, for instance, with carbon dioxide, they have a significant greenhouse effect. It is estimated that due to the CFCs now present in the atmosphere the radiative forcing in the tropics (at higher latitudes there is a compensating effect due to ozone reduction which is explained below) is about 0.25 Wm^{-2} – or about 20 per cent of the radiative forcing due to all greenhouse gases. This forcing will only decrease very slowly next century.

Turning now to ozone, the effect from ozone depletion is complex because the amount by which ozone greenhouse warming is reduced depends critically on the height in the atmosphere at which it is being destroyed. Further, ozone depletion is concentrated at high latitudes while the greenhouse effect of the CFCs is uniformly spread over the globe. In tropical regions there is virtually no ozone depletion so no change in the ozone greenhouse effect. At mid-latitudes, very approximately, the greenhouse effects of ozone reduction and of the CFCs compensate for each other. In polar regions the reduction in the greenhouse effect of ozone more than compensates for the greenhouse warming effect of the CFCs[21].

As CFCs are phased out, they are being replaced to some degree by other halocarbons – hydrochloro-fluorocarbons (HCFCs) and hydrofluorocarbons (HFCs). In Copenhagen in 1992, the international community decided that HCFCs would also be phased out by the year 2030. While being less destructive to ozone than the CFCs, they are still greenhouse gases. The HFCs contain no chlorine or bromine, so they do not destroy ozone and are not covered by the Montreal Protocol. Because of their shorter lifetime, typically tens rather than hundreds of years, the concentration in the atmosphere of both the HCFCs and the HFCs, and therefore their contribution to global warming for a given rate of emission, will be less than for the CFCs. However, should the rate of their manufacture threaten to increase substantially their potential contribution to greenhouse warming must be exposed and properly considered.

Concern has also recently been extended to some other related compounds which are greenhouse gases, the perfluorocarbons (eg CF_4, C_2F_6) and sulphur hexafluoride (SF_6), which are produced in some industrial processes. Because they possess very long atmospheric lifetimes, probably more than 1000 years, all emissions of these gases accumulate in the atmosphere and will continue to influence climate for thousands of years. If emissions of these gases increase substantially from their present low levels, their effect on the climate could become significant.

Ozone is also present in the lower atmosphere or troposphere, where some of it is transferred downwards from the stratosphere and where some is generated by chemical action, particularly as a result of the action of sunlight on the oxides of nitrogen. It is especially noticeable in polluted atmospheres near the surface; if present in high enough concentration, it can become a health hazard.

It is also generated at levels in the upper troposphere as a result of the nitrogen oxides emitted from aircraft exhausts. In the northern hemisphere the limited observations available together with model simulations of the chemical reactions leading to ozone formation suggest that ozone concentrations in the troposphere may have doubled since pre-industrial times – an increase which is estimated to have led to a global average radiative forcing of between 0.2 and 0.6 Wm^{-2}. Estimates of the influence on radiative forcing of the additional ozone arising from the flights of aircraft in the upper troposphere suggest that it may be significant, but unlikely to exceed that of the carbon dioxide emitted from the combustion of aviation fuel which is about 3 per cent of current fossil fuel consumption.

Gases with an indirect greenhouse effect

I have described all the gases present in the atmosphere which have a direct greenhouse effect. There are also gases which through their chemical action on greenhouse gases, for instance on methane or on lower atmospheric ozone, have an influence on the overall size of greenhouse warming. Carbon monoxide (CO) and the nitrogen oxides (NO and NO_2) emitted, for instance, by motor vehicles, are some of these. Carbon monoxide has no direct greenhouse effect of its own but, as a result of chemical reactions, it forms carbon dioxide. These reactions also affect the amount of the hydroxyl radical (OH) which in turn affects the concentration of methane.

Much research needs to be done on many of the chemical processes in the atmosphere in order to elucidate the details of these processes which have indirect effects on greenhouse gases. While it is important to do this and to take them properly into account, it is also important to recognize that their combined effect is much less than that of the major contributors to human-generated greenhouse warming, namely carbon dioxide and methane.

Particles in the atmosphere

Small particles suspended in the atmosphere (often known as *aerosol*) affect its energy balance because they both absorb radiation from the sun and scatter it back to space. We can easily see the effect of this on a bright day in the summer with a light wind when downwind of an industrial area. Although no cloud appears to be present, the sun appears hazy. We call it 'industrial haze'. Under these conditions a significant proportion of the sunlight incident at the top of the atmosphere is being lost as it is scattered back and out of the atmosphere by the millions of small particles (typically between 0.001 and 0.01 mm in diameter) in the haze.

Atmospheric particles come from a variety of sources. They arise partially from natural causes; they are blown off the land surface, especially in desert areas; they result from forest fires and they come from sea spray. From time to time large quantities of particles are injected into the upper atmosphere from volcanoes – the Pinatubo volcano which erupted in 1991 provides a good example (see Chapter 5). Some particles are also formed in the atmosphere itself, for instance sulphate particles from the sulphur-containing gases emitted from volcanoes.

Other particles arise from human activities – from biomass burning (e.g. the burning of forests) and the sulphates and soot resulting from the burning of fossil fuels. The sulphate particles are particularly important. They are formed as a result of chemical action on sulphur dioxide, a gas which is produced in large quantities by power stations and other industries in which coal and oil (both of which contain sulphur in varying quantities) are burnt. Because these particles remain in the atmosphere only for about 5 days on average, their effect is mainly confined to regions near the sources of the particles, i.e. the major industrial regions of the northern hemisphere (Fig. 3.7). Over limited regions of the northern hemisphere the radiative effect of these particles is comparable in size, although opposite in effect, to that of human-generated greenhouse gases up to the present time.

Averaged over the globe, the amount of sunlight scattered back to space by the particles from human-generated sources – the sulphates and the particles from biomass burning – has been estimated to amount to a loss of about 0.6 Wm^{-2}. Absorption of sunlight by soot particles amounts to perhaps an addition of 0.1 Wm^{-2} so that the net global average direct radiative forcing is estimated as –0.5 Wm^{-2}, with a range of uncertainty of a factor of two either way.

The above has been described as *direct* radiative forcing. There is a further way by which particles in the atmosphere could influence the climate; that is through their effect on cloud formation which we describe as *indirect* radiative forcing and which arises as follows. If particles are present in large numbers when clouds are forming, the resultant cloud will consist of a large number of smaller drops – smaller than would otherwise be the case – this is similar to what happens as polluted fogs form in cities. Such a cloud will be more highly reflecting to sunlight than one consisting of larger particles, thus further increasing the energy loss resulting from the presence of the particles. It is not well known how large this effect is likely to be; estimates of its size, when globally averaged, fall in the range of an energy loss of between 0 and 1.5 Wm^{-2}. To refine these estimates, more studies need to be made especially by making careful measurements on suitable clouds.

Fig 3.7
Modelled geographic distribution of estimates of the annual mean direct radiative forcing (in watts per square metre) from anthropogenic sulphate aerosols in the troposphere[22]. The negative radiative forcing is largest over regions close to

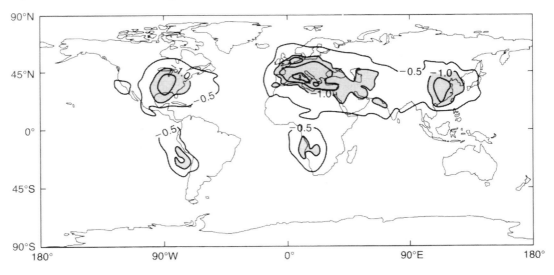

The estimates for the radiative effects of particles can be compared with the global average radiative forcing to date due to the increase in greenhouse gases of about 2.6 Wm^{-2}. Comparing global average forcings, however, is not the whole story. Because of the large regional variation of particles in the atmosphere, any effect they have on the climate can be expected to be substantially different from the effect of increases in greenhouse gases which is much more uniform over the globe. This will be illustrated in Chapter 5 where we consider the likely pattern of anthropogenic climate change to date; more consideration of it will also be given in Chapter 6 where it wil be shown that projections of future climate change can depend critically on what assumptions are made about the likely concentrations of atmospheric particles in the future.

An important factor which will influence the future concentrations especially of sulphate particles is 'acid rain' pollution which is mainly caused by sulphur dioxide emissions. This leads to the degradation of forests and fish stocks in lakes especially in regions downwind of major industrial areas. Serious efforts are therefore underway, especially in Europe and North America, to curb these emissions to a substantial degree. Although the amount of sulphur-rich coal being burnt elsewhere in the world, for instance in Asia, is increasing rapidly, the damaging effects of sulphur pollution are such that tight controls on sulphur emissions are likely to be extended to these regions also. For the globe as a whole therefore, sulphur emissions are likely to rise much less rapidly than emissions of carbon dioxide. The climate change resulting from an increase in sulphate particles therefore will become increasingly less by comparison with that from the likely increase of greenhouse gases.

Estimates of radiative forcing

This chapter has summarized current scientific knowledge about the sources and sinks of the main greenhouse gases and the exchanges which occur between the components of the climate system – the atmosphere, the ocean and the land surface- including the close balances which are maintained between the different components and the way in which these balances are being disturbed by human-generated emissions. Different assumptions about future emissions have been used to generate emission scenarios. From these scenarios estimates have been made (for carbon dioxide, for instance, using a computer model of the carbon cycle) of likely increases in greenhouse gas concentrations in the future.

Given information about the possible increases in greenhouse gases, the next step is to calculate the effect of these increases on the amounts of thermal (infrared) radiation absorbed and emitted by the atmosphere. This is done using information about how the different gases absorb radiation in the infrared part of the spectrum, as mentioned in Chapter 2. The radiative forcing associated with the increases in each of the gases can then be calculated.

In Figs. 3.8 and 3.9 are brought together the estimates of global average radiative forcing for the different greenhouse gases and for tropospheric aerosols of different origins which we have been considering in this chapter. The radiative forcings from 1850 to date are shown in Fig. 3.8 and as projected

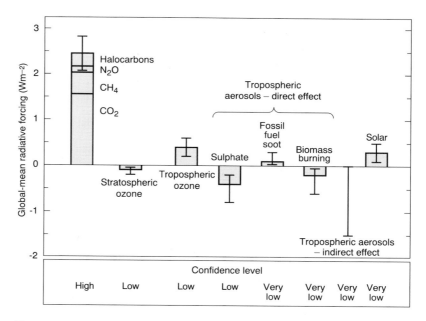

Fig 3.8
Estimates of the globally and annually averaged anthropogenic radiative forcings (in watts per square metre) due to changes in concentrations of greenhouse gases and aerosols from pre-industrial times to the present (1992) and to natural changes in solar output from 1850 to the present[23]. The height of the rectangular bars indicates mid-range estimates of the forcings; the error bars show estimates of the uncertainty ranges. The contributions of individual gases to the direct greenhouse forcing are indicated in the first bar; the greenhouse forcings associated with changes in ozone are shown in the second and third bars. Estimates of the forcings (globally averaged values) due to changes in aerosols are shown in the following bars; note that only the range of values, not a central estimate is shown for the indirect aerosol forcing, because of its large uncertainty.

for next century under the business-as-usual emission scenario IS 92a in Fig. 3.9. (for assumptions underlying IS 92a see the table on p. 32).

When considering radiative forcing of climate, the question is bound to be asked as to whether variations in the energy output of the sun have occurred or are likely to occur which could affect the climate. We shall see, for instance, in the next chapter that ice ages in the past have been triggered by variations in the geometry of the Earth's orbit. It is considered possible that the sun's output itself could vary by small amounts over time (see box in Chapter 6). Figure 3.8 indicates the range of values of solar variability which it is estimated may have occurred during the last century showing that its influence is much less than that of the increase in greenhouse gases.

It is often convenient to make calculations of the greenhouse effect assuming that carbon dioxide is the only greenhouse gas. Because of this it is also useful to convert the other greenhouse gases to equivalent amounts of carbon dioxide, in other words to the amounts of carbon dioxide which would give the same radiative forcing. The information in Fig. 3.9 enables the conversion to be carried out[24]. For instance, the increases in all the greenhouse gases to date

Fig 3.9
Radiative forcing[25]: (a) Radiative forcing
components resulting from the IS 92a emission
scenario for 1990 to 2100. The 'Total non-CO_2
trace gases' curve includes the radiative forcing
from methane (including methane-related
increases in stratospheric water vapour),
nitrous oxide, tropospheric ozone and the
halocarbons (including the negative forcing
effect of stratospheric ozone depletion) – more
details in (b). The three aerosol components
are: direct sulphate, indirect sulphate and direct
biomass burning.
(b) Non-CO_2 trace gas radiative forcing
components. 'Cl/Br direct' is the direct radiative
forcing from chlorine and bromine containing
halocarbons; emissions of these under IS 92a
have been modified to take account of the
Montreal Protocol and its Adjustments and
Amendments. The indirect forcing from these
compounds (through stratospheric ozone
depletion) is shown separately (Strat O_3). The
tropospheric ozone forcing takes account of
concentration changes due only to the indirect
effect of methane.

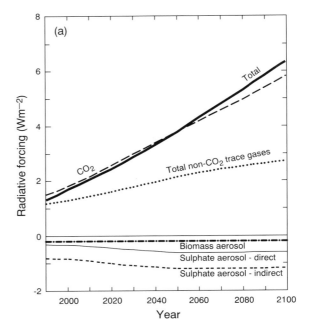

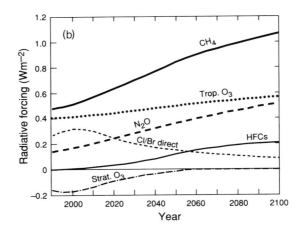

are equivalent to an increase in carbon dioxide of about 50 per cent and, under
the IS 92a scenario, doubling of the equivalent carbon dioxide amount from
pre-industrial times will occur in about 2030. Chapters 5 and 6 will explain
how these estimates of radiative forcing can be incorporated into computer climate models to predict the future climate change which is likely to occur
because of human activities. However, before considering predictions of future
climate change, it is helpful to gain perspective by looking at some of the climate changes which have occurred in the past.

Questions 1   The lifetime of a carbon dioxide molecule in the atmosphere before it is exchanged with the ocean is typically less than a year, while the time taken for an increase in carbon dioxide concentration from fossil fuel burning to diminish substantially is typically many years. Explain the reasons for this difference.

2   Estimate how much carbon dioxide you emit each year through breathing.

3   Estimate the size of your share of carbon dioxide emissions from the burning of fossil fuels.

4   A typical city in the developed world with a population of about one million produces about half a million tonnes of municipal waste each year. Suppose the waste is buried in a landfill site where the waste decays producing equal quantities of carbon dioxide and methane. Making assumptions about the likely carbon content of the waste and the proportion which eventually decays, estimate the annual production of methane. If all the methane leaks away, using the information in Note 17, compare the greenhouse effect of the carbon dioxide and methane produced from the landfill site with that of the carbon dioxide produced if the waste were incinerated instead.

5   A new forest is planted containing a million trees which will mature in 40 years. Estimate the amount of carbon sequestered per year by the forest.

6   Find figures for the amount of fuel used by a typical aircraft and the size of fleets of the world's airlines and airforces and estimate the carbon dioxide emitted globally each year by the world's aircraft.

7   Search for information about the ozone hole and explain why it occurs mainly in the Antarctic.

8   What are the main uses of CFCs? Suggest ways in which the emissions of CFCs to the atmosphere could be reduced more rapidly.

9   Evidence is sometimes presented suggesting that the variations in global average temperature over the last century or more can all be explained as due to variations in the energy output of the sun. There is therefore nothing left to attribute to the increase in greenhouse gases. What is the fallacy in this argument?

10   With the use of the formula in Note 25, calculate the radiative forcing due to carbon dioxide for atmospheric concentrations of 150, 280, 450, 560 and 1000 ppmv.

Notes

1   It is convenient to define radiative forcing as the radiative imbalance at the top of the troposphere rather than at the top of the whole atmosphere.

2   Adapted from S. H. Schneider 'The changing climate' *Scien. Amer.*, September 1989, pp. 38–47 and from U. Siegenthaler and J. L. Sarmiento, 'Atmospheric carbon dioxide and the ocean', *Nature*, **365**, 1993, pp. 119–25.

3   From R. T. Watson *et al.*, 'Greenhouse gases and aerosols' in *Climate Change, the IPCC Scientific Assessment*, eds. J. T. Houghton, G. J. Jenkins and J. J. Ephraums, CUP, 1990, pp. 1–40 (updated).

4   The data for diagram (b) are from C. D. Keeling of the Scripps Institute of Oceanography and P. Tans of the National Oceanic and Atmospheric Administration of the United States.

5 From *Climate Change 1995: The Science of Climate Change*, eds. J. T. Houghton, L. G. Meira Filho, B. A. Callender, N. Harris, A. Kattenberg and K. Maskell, CUP, 1996, Technical Summary.

6 From *Climate Change 1995: The Science of Climate Change*, eds. J. T. Houghton, L. G. Meira Filho, B. A. Callender, N. Harris, A. Kattenberg and K. Maskell, CUP, 1996, Technical Summary.

7 Both (a) and (b) are from D. Schimel *et al.*, 'CO$_2$ and the carbon cycle' chapter 1 in *Climate Change* 1994 published for the IPCC by CUP, 1995.

8 More details in J. M. Melillo *et al*, chapter 9, *Climate Change 1995: The Science of Climate Change*, eds. J. T. Houghton, L. G. Meira Filho, B. A. Callender, N. Harris, A. Kattenberg and K. Maskell, CUP, 1996.

9 J. Woods and W. Barkmann, 'The plankton multiplier – positive feedback in the greenhouse', *J. of Plankton Research*, **15**, 1993, pp. 1053–74.

10 R. F. Keeling, S. C. Piper and M. Heimann, 'Global and hemispheric sinks deduced from changes in atmospheric O$_2$ concentration' *Nature*, **381**, pp. 218–21, 1996.

11 J. Leggett, W. J. Pepper and R. J. Swart, 'Emissions Scenarios for the IPCC: an update' in *Climate Change 1992, The Supplementary Report to the IPCC Scientific Assessment*, eds. J. T. Houghton, B. A. Callander and S. K. Varney, CUP, 1992, pp. 69–95.

12 World Energy Council Commission Report, *Energy for tomorrow's world*, World Energy Council, London, 1993.

13 These numbers are from a World Bank estimate; they are close to the UN Medium Case.

14 From World Energy Council Commission Report, *Energy for tomorrow's world*, World Energy Council, London, 1993, Appendix E, pp. 30–306. The model used in these calculations was developed by Professor T. Wigley. For the calculations in the 1995 IPCC Report the Bern model was used which gives slightly higher values, for instance, about 710 ppmv in 2100 for IS 92a scenario.

15 For more information on the model see T. M. C. Wigley, 'Balancing the carbon budget: implications for projections of future carbon dioxide changes', *Tellus*, **45B**, 1993, pp. 409–25.

16 In 1992 the increase slowed to almost zero. The reason for this is not known but one suggestion is that, because of recent changes in Russia, the leakage from Siberian natural gas pipelines has been much reduced.

17 The ratio of the enhanced greenhouse effect of a molecule of methane compared to a molecule of carbon dioxide is known as its global warming potential (GWP). The figure of about 7.5 given here for the GWP of methane is for a time horizon of 100 years – see J. Lelieveld and P. J. Crutzen, *Nature*, **355**, 1992, pp. 339–41; see also D. Schimel *et al.*, chapter 2, *Climate Change 1995: The Science of Climate Change*, eds. J. T. Houghton, L. G. Meira Filho, B. A. Callender, N. Harris, A. Kattenberg and K. Maskell, CUP, 1996. About half of the contribution of methane to the greenhouse effect is because of its direct effect on the outgoing thermal radiation. The other half arises because of its influence on the overall chemistry of the atmosphere. Increased methane eventually results in small increases in water vapour in the upper atmosphere, in tropospheric ozone and in carbon dioxide, all of which in turn add to the greenhouse effect.

18 The effective lifetime of methane in the atmosphere is complex because it depends on the methane concentration. This is because the concentration of the radical OH (interaction with which is the main cause of methane destruction), because of chemical feedbacks, is itself dependent on the methane concentration.

19 From R. T. Watson *et al.*, *Climate Change, the IPCC Scientific Assessment*, eds. J. T. Houghton, G. J. Jenkins and J. J. Ephraums, CUP, 1990, pp. 1–40 (updated).

20 From M. Prather *et al.*, 'Other trace gases and atmospheric chemistry', chapter 2 in *Climate Change 1994* published for the IPCC by CUP, 1995; and D. Schimel *et al.*, chapter 2, in *Climate Change 1995: The Science of Climate Change*, eds. J. T. Houghton, L. G. Meira Filho, B. A. Callender, N. Harris, A. Kattenberg and K. Maskell, CUP, 1996. The figure for atmospheric increase is an average from 1980 to 1990.

21 More detail on this and the radiative effects of minor gases and particles can be found in I. S. A. Isaksen *et al.*, 'Radiative Forcing of Climate' in *Climate Change 1992, The Supplementary Report to the IPCC Scientific Assessment*, eds. J. T. Houghton, B. A. Callander and S. K. Varney, CUP, 1992, pp. 47–67.

22 After J. T. Kiehl and B. P. Briegleb, from Summary for Policymakers, in *Climate Change 1994* published for the IPCC by CUP, 1995.

23 From *Climate Change 1995: The Science of Climate Change*, eds. J. T. Houghton, L. G. Meira Filho, B. A. Callender, N. Harris, A. Kattenberg and K. Maskell, CUP, 1996, Technical Summary.

24 The assumption that greenhouse gases may be treated as equivalent to each other is a good one for many purposes. However, because of the differences in their radiative properties, accurate modelling of their effect should treat them separately. More details of this problem are given in W. L. Gates *et al.*, 'Climate modelling, climate prediction and model validation' in *Climate Change 1992, The Supplementary Report to the IPCC Scientific Assessment*, eds. J. T. Houghton, B. A. Callander and S. K. Varney, CUP, 1992, pp. 171–5.

25 From *Climate Change 1995: The Science of Climate Change*, eds. J. T. Houghton, L. G. Meira Filho, B. A. Callender, N. Harris, A. Kattenberg and K. Maskell, CUP, 1996, Technical Summary. A useful formula for the radiative forcing R from atmospheric carbon dioxide when its atmospheric concentration is C ppmv is: $R = 6.3 \ln (C/C_0)$ where C_0 is its pre-industrial concentration of 280 ppmv.

4 Climates of the Past

To obtain some perspective against which to view future climate change, it is helpful to look at some of the climate changes which have occurred in the past. This chapter will briefly consider climatic records and climate changes in three periods: the last hundred years, then the last thousand years and finally the last million years. At the end of the chapter some interesting recent evidence for the existence of relatively rapid climate change at various times during the past one or two hundred thousand years will be presented.

The last hundred years

The 1980s and early 1990s have brought some unusually warm years for the globe as a whole (see Chapter 1); eight of the nine warmest years during the past century have occurred during this period (up to and including 1995). 'Warmest' means only a few tenths of a degree Celsius at most, but in terms of the global average such differences are quite significant. Figure 4.1 shows the global average temperature since 1860; an increase over this period has taken place of between 0.3 °C and 0.6 °C. Although there is a distinct trend, the increase is by no means a uniform one; in fact, some periods of cooling as well as warming have occurred.

A sceptic may wonder how a diagram like Fig. 4.1 can be prepared and whether any reliance can be placed upon it. After all, temperature varies from place to place, from season to season and from day to day by many tens of

Fig. 4.1 Changes in global average surface temperature (surface air temperature over land and sea surface temperatures combined) from 1860 to 1995 relative to the period 1961–90[1]. The solid curve represents smoothing of the annual values to suppress sub-decadal timescale variations.

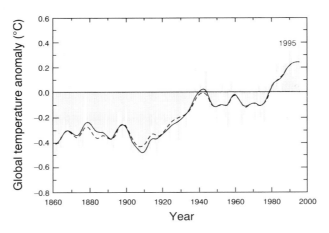

degrees. How,then, can a global average changing by a few tenths of a degree have any meaning?

First of all, just how is a change in global average temperature estimated from a combination of records of changes in the near surface temperature over land and changes in the temperature of the sea surface? To estimate the changes over land, weather stations are chosen where consistent observations have been taken from the same location over a substantial proportion of the whole 130 year period. Changes in sea surface temperature have been estimated by processing over sixty million observations from ships – mostly merchant ships – over the same period. All the observations, from land stations and from ships, are then located within a grid of squares, say 1° of latitude by 1° of longitude, covering the Earth's surface. Observations within each square are averaged; the global average is obtained by averaging (after weighting them by area) over the averages for each of the squares.

A number of research groups in different countries have made careful and independent analyses of these observations. In somewhat different ways they have made allowances for factors which could have introduced artificial changes in the records. For instance, the record at some land stations could have been affected by changes in their surroundings as these have become more urban. In the case of ships, the standard method of observation used to be to insert a thermometer into a bucket of water taken from the sea. Small changes of temperature have been shown to occur during this process; the size of the changes varies between day and night and is also dependent on several other factors including the material from which the bucket is made – over the years wooden, canvas and metal buckets have been variously employed. Nowadays, a large proportion of the observations are made by measuring the temperature of the water entering the engine cooling system. Careful analysis of the effects of these details on observations both on land and from ships has enabled appropriate corrections to be made to the record, and good agreement has been achieved between analyses carried out at different centres.

Confidence that the observed variations are real is increased by noticing that the trend and the shape of the changes are similar when different selections of the total observations are made. For instance, the separate records from the land and sea surface and from the northern and southern hemispheres are closely in accord. Further indirect indicators such as changes in borehole temperatures and sub-surface ocean temperatures, decrease in snow cover and glacier shrinkage provide independent support for the observed warming.

During the last twenty years or so observations have been available from satellites orbiting around the Earth. Their great advantage is that they automatically provide data with global coverage, which is often lacking in other data sets. The length of the record from satellites, however, is generally less than twenty years, a comparatively short period in climate terms. It has been suggested that satellite measurements of lower atmospheric temperature since 1979 are not consistent with the trend of rising temperatures in surface observations; a careful analysis, however, confirms that there is satisfactory agreement between the satellite and the surface measurements (see box).

The most obvious feature of the climate record illustrated in Fig. 4.1 is that of considerable variability, not just from year to year, but from decade to

Atmospheric temperature observed by satellites

Since 1979 meteorological satellites flown by the National Oceanic and Atmospheric Administration (NOAA) of the United States have carried a microwave instrument, the Microwave Sounding Unit (MSU), for the remote observation of the average temperature of the lower part of the atmosphere. Figure. 4.2 shows the record of global average temperature deduced from the MSU and compares it with data from balloon sondes and surface data for the same period. All three measurements, although they are from very different sources and do not apply to precisely the same part of the atmosphere, show good agreement as they track significant variations during the period. The plots also illustrate the difficulty of deriving accurate trends from such a short period of record where there is also evident such substantial variability. Comparisons have, however, been made of the trends from the three sources after appropriate corrections have been made for the known effects of the Pinatubo volcano and El Niño events. When this has been done, the trends over the period in global average temperature per decade are 0.09 °C from the MSU data, 0.10 °C from the radiosonde data and 0.17 °C from the surface air temperature data – which can be considered reasonable agreement especially when it is realised that there are reasons for expecting, especially over land, a different response at the surface from that in the mid-troposphere.

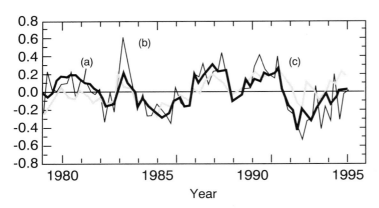

Fig. 4.2
Comparison, for 1979 to 1994, of global average temperature (shown as deviations from the 1979-1994 average) from (a) the MSU channel covering the lower atmosphere (approx 1-10 km in altitude), (b) balloon borne radiosondes, averaged over the lower atmosphere, (c) surface air temperature data as Fig. 4.1[2].

decade. Some of this variability will have arisen through causes external to the atmosphere and the oceans, for instance, as a result of volcanic eruptions such as those of Krakatoa in 1883 or of Pinatubo in the Philippines in 1991 (the low global average temperature in 1992, compared with 1990, 1991, 1994 and 1995 is almost certainly due to the Pinatubo volcano). But there is no need to invoke volcanoes or other external causes to explain most of the variations in the record. Their most likely cause is internal variations within the total climate system, for instance between different parts of the ocean[3].

The warming during the 20th century has not been uniform over the globe. The recent warming has been greatest over the mid latitude continents in winter and spring – some of the most detailed studies have been for the United

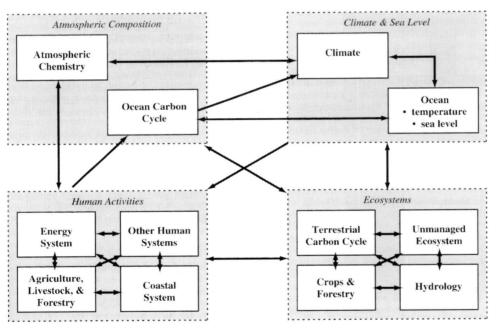

The ocean's deep circulation

For climate change over periods up to a decade, only the upper layers of the ocean have any substantial interaction with the atmosphere. For longer periods, however, links with the deep ocean circulation become important. The effects of changes in the deep circulation are of particular importance.

Experiments using chemical tracers, for instance those illustrated in Fig. 5.20 (see next box) have been helpful in indicating the regions where strong coupling to the deep ocean occurs. To sink to the deep ocean, water needs to be particularly dense, in other words both cold and salty. There are two main regions where such dense water sinks down to the deep ocean, namely in the north Atlantic Ocean between Scandinavia and Greenland and in the region of Antarctica. Salt-laden deep water formed in this way contributes to a deep ocean circulation that involves all the oceans (Fig. 5.18).

An important link exists between this deep ocean circulation and the hydrological (water) cycle. Suppose, for instance, that there is persistent increased precipitation in the north Atlantic region. The surface water will become less salty and therefore less dense. It will not sink so readily and the deep water formation will be inhibited. This sequence of events has in fact been reported in runs of the climate model at the Geophysics Fluid Dynamics Research Laboratory at Princeton in the USA. As the amount of carbon dioxide in the model was increased, more precipitation occurred over the north Atlantic. With doubled carbon dioxide the rate of deep water formation was reduced by 30 per cent; with quadrupled carbon dioxide, it ceased altogether bringing large changes of climate to that region[23].

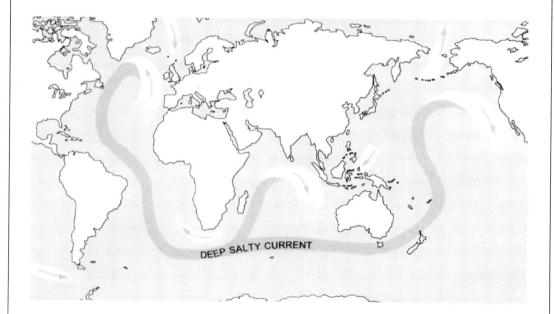

Fig. 5.18
Deep water formation and circulation. The deep salty current originates in the north Atlantic. Northward flowing water near the surface which is unusually salty becomes cooler and even more salty through evaporation, so increasing its density and causing it to sink[24].

instance through comparisons between the simulation and observation of the movement of chemical tracers (see box on page opposite).

Once a comprehensive climate model has been formulated it can be tested in three main ways. Firstly, it can be run for a number of years of simulated time and the climate generated by the model compared in detail to the current climate. For the model to be seen as a valid one, the average distribution and the seasonal variations of appropriate parameters such as surface pressure, temperature and rainfall have to compare well with observation. In the same way, the variability of the model's climate must be similar to the observed variability. Climate models which are currently employed for climate prediction stand up well to such comparisons[25].

Secondly, models can be compared against simulations of past climates when the distribution of key variables was substantially different than at present, for example the period around 9,000 years ago when the configuration of the Earth's orbit around the sun was different (see Fig. 5.19). The perihelion (minimum Earth-sun distance) was in July rather than in January as it is now; also the tilt of the Earth's axis was slightly different from its current value (24° rather than 23.5°). Resulting from these orbital differences (see Chapter 4), there were significant differences in the distribution of solar radiation throughout the year. The incoming solar energy when averaged over the northern hemisphere was about 7 per cent greater in July and correspondingly less in January.

When these altered parameters are incorporated into a model, a different climate results. For instance, northern continents are warmer in summer and colder in winter. In summer a significantly expanded low pressure region develops over north Africa and south Asia because of the increased land-ocean temperature contrast. The summer monsoons in these regions are strengthened and there is increased rainfall. These simulated changes are in qualitative agreement with paleoclimate data; for example, these data provide evidence for that period (around 9,000 years ago) of lakes and vegetation in the southern Sahara about 1,000 km north of the present limits of vegetation.

Fig. 5.19
Changes in the Earth's elliptical orbit from the present configuration to 9,000 years ago and (bottom) changes in the average solar radiation during the year over the northern hemisphere[26].

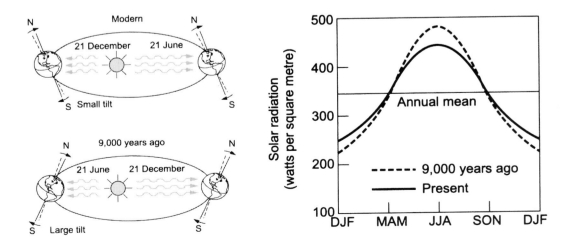

Modelling of tracers in the ocean

A test which assists in validating the ocean component of the model is to compare the distribution of a chemical tracer as observed and as simulated by the model. In the 1950s radioactive tritium (an isotope of hydrogen) released in the major atomic bomb tests entered the oceans and was distributed by the ocean circulation and by mixing. Figure 5.20 shows good agreement between the observed distribution of tritium (in tritium units) in a sec-

tion of the western north Atlantic Ocean about a decade after the major bomb tests and the distribution as simulated by a 12-level ocean model. Similar comparisons have been made more recently of the measured uptake of one of the freons CFC-11, whose emissions into the atmosphere have increased rapidly since the 1950s, compared with the modelled uptake.

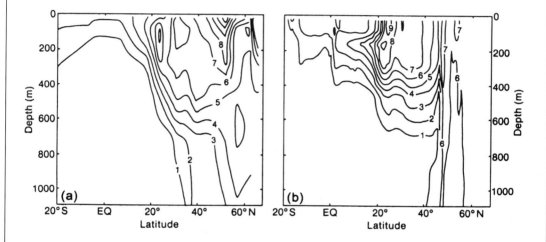

Fig. 5.20
The tritium distribution in a section of the western north Atlantic Ocean approximately one decade after the major bomb tests; as observed in the GEOSECS programme (a) and as modelled (b)[27].

Fig 5.21
The predicted and observed changes in global land and ocean surface air temperature after the eruption of Mount Pinatubo, in terms of three-month running averages from April to June 1991 to March to May 1995[28].

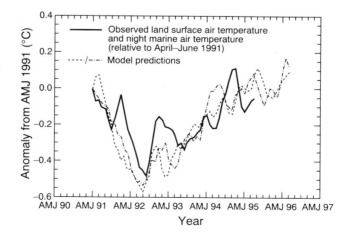

The accuracy and the coverage of data available for these past periods are limited. However, the model simulations for 9,000 years ago, described above, and those for other periods in the past have demonstrated the value of such studies in the validation of climate models.

A third way in which models can be validated is to use them to predict the effect of large perturbations on the climate. Good progress is being achieved with the prediction of El Niño events and the associated climate anomalies up to a year ahead (see earlier in the chapter). Other short-term perturbations are due to volcanic eruptions, the effects of which were mentioned in Chapter 1. Several climate models have been run in which the amount of incoming solar radiation has been modified to allow for the effect of the volcanic dust from Mount Pinatubo, which erupted in 1991 (Fig. 5.21). Successful simulation of some of the regional anomalies of climate which followed that eruption, for instance, the unusually cold winters in the Middle East and the mild winters in western Europe, has also been achieved by the models[29].

In these three ways, which cover a range of timescales, confidence has been built up in the ability of models to predict climate change due to human activities.

Comparison with observations

Fourteen centres in the world located in ten countries are currently running climate models of the kind we have described in which the circulations of the atmosphere and the ocean are fully coupled together. Some of these models have been employed to simulate the climate of the past century allowing for the increases in the concentrations of greenhouse gases and aerosols.

An example of such a simulation is shown in Fig. 5.22 where the observed record of global average surface air temperature is compared with model simulations taking into account the increase in greenhouse gases only and greenhouse gases and sulphate aerosols combined. Three interesting features of Fig. 5.22 can be noted. Firstly, that the agreement between the model

Fig. 5.22
Simulated global annual surface air temperature from 1860 to 1990 allowing for increases in greenhouse gases only (dashed curve) and greenhouse gases and sulphate aerosols combined (full curve) compared with observed changes over the same period[30].

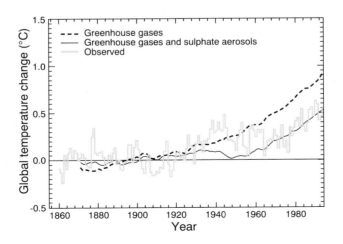

simulation and observations is much better for the curve which includes the cooling effect of increased aerosols in addition to the warming effect of increased greenhouse gases. Secondly, the model simulation shows variability up to a tenth of a degree Celsius or more over periods of a few years up to decades. This variability is due to internal exchanges in the model between different parts of the climate system, and is not dissimilar to that which appears in the observed record. Thirdly, due to the slowing effect of the oceans on climate change, the warming observed or modelled so far is less than would be expected if the climate system were in equilibrium under the amount of radiative forcing due to the current increase in greenhouse gases and aerosols.

The question, however, arises as to whether a diagram such as Fig. 5.22 provides strong evidence that global warming due to the increase in greenhouse gases has actually been observed in the climate record. In other words has the 'signal' which can be attributed to global warming risen sufficiently above the 'noise' of natural variability?

The IPCC Report published in 1990[31] made a carefully worded statement to the effect that, although the size of the observed warming is broadly consistent with the predictions of climate models, it is also of similar magnitude to natural climate variability. An unequivocal statement that anthropogenic climate change had been detected could not therefore be made at that time.

Since 1990 many more studies have been carried out. We mentioned in Chapter 4 the studies which have compared the global average temperature over the last hundred years with what is known of climate variability on that timescale. The studies concluded that the observed rise in temperature is such that it is unlikely to be entirely natural in origin. The example shown in Fig. 5.22 suggests that the 'signal' which appears to be rising above the 'noise' might be attributed to the recent increase in greenhouse gases. More particularly, studies comparing the modelled patterns of temperature change both in their geographical distribution and their distribution in the vertical show significant correlations with observed temperature changes over the last 30 years (Fig. 5.23). The IPCC 1995 Report[32] therefore reached the cautious conclusion which is worth stating in full.

Our ability to quantify the human influence on global climate is currently limited because the expected signal is still emerging from the noise of natural climate variability, and because there are uncertainties in key factors. These include the magnitude and patterns of long term natural variability and the time-evolving pattern of forcing by, and response to, changes in the concentrations of greenhouse gases and aerosols, and land surface changes. Nevertheless, the balance of evidence suggests a discernible influence on global climate.

Confidence having been established in climate models in the ways we have outlined in the last two sections, these models can now be used to generate projections of the likely climate change in the future due to human activities. Details of such projections will be presented in the next chapter.

Fig 5.23
Modelled and observed changes in the zonal-mean, annual average temperature structure of the atmosphere from an altitude of about 1.5 km where the pressure is 850 hPa to about 20 km altitude where the pressure is 50 hPa[33]. The modelled changes are (a) for the increase in greenhouse gases from preindustrial levels, (b) as in (a) but including also the effect of sulphate aerosols. (c) shows observed changes as measured by radiosondes over the period 1963 to 1988. A common pattern of stratospheric cooling and tropospheric warming is evident in the observations and both model experiments, but better agreement obtains when the effects of aerosols are included in the model. The more pronounced stratospheric cooling in the observations is due at least in part to the reduction in ozone which has also occurred in this region during the period; this reduction in ozone was not included in the model calculation.

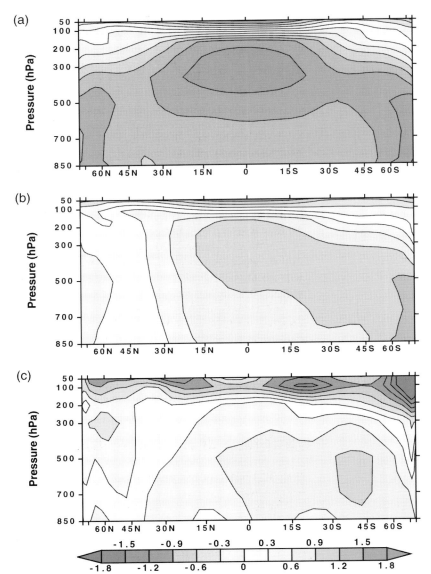

Is the climate chaotic?

Throughout this chapter the implicit assumption has been made that climate change is predictable and that models can be used to provide predictions of climate change due to human activities. Before leaving this chapter I want to consider whether this assumption is justified.

The capability of the models themselves has been demonstrated so far as weather forecasting is concerned. They also possess some skill in seasonal forecasting. They can provide a good description of the current climate and its seasonal variations. Further, they provide predictions which on the whole are reproducible and which are reasonably consistent between different models. But, it might be argued, this consistency could be a property of the models rather than of the climate. Is there any other evidence to support the view that the climate is predictable?

A good place to look for further evidence is in the record of climates of the past, presented in Chapter 4. Correlation between the Milankovitch cycles in the Earth's orbital parameters and the cycles of climate change over the past half million years (see Fig. 4.4, also Fig. 5.19) provides strong evidence to substantiate the Earth's orbital variations as the main factor responsible for the triggering of climate change. The nature of the feedbacks which control the very different amplitudes of response to the three orbital variations still need to be understood. Some 60 ± 10 per cent of the variance in the record of global average temperature from paleontological sources over the past million years occurs close to frequencies identified in the Milankovitch theory. The existence of this surprising amount of regularity suggests that the climate system is not strongly chaotic so far as these large changes are concerned, but responds in a largely predictable way to the Milankovitch forcing.

This Milankovitch forcing arises from changes in the distribution of solar radiation over the Earth because of variations in the Earth's orbit. Changes in climate as a result of the increase of greenhouse gases are also driven by changes in the radiative regime at the top of the atmosphere. These changes are not dissimilar in kind (although different in distribution) from the changes which provide the Milankovitch forcing. It can be argued therefore that the increases in greenhouse gases will also result in a largely predictable response.

The future of climate modelling

Very little has been said in this chapter about the biosphere. Chapter 3 referred to comparatively simple models of the carbon cycle which include chemical and biological processes and simple non-interactive descriptions of atmospheric processes and ocean transport. The large three-dimensional global circulation climate models described in this chapter contain a lot of dynamics and physics but no interactive chemistry or biology. Detailed scientific knowledge of these areas is not yet sufficient to warrant their inclusion in such complex models; further their inclusion would, at the present, be too demanding in computer time. Eventually, however, it will be important and necessary to have models which are fully comprehensive and interactive and which include chemical and biological processes in the atmosphere, the ocean and on the land.

Climate modelling continues to be a rapidly growing science. Although useful attempts at simple climate models were made with early computers it is only during the last ten years or so that computers have been powerful enough for coupled atmosphere-ocean models to be employed for climate prediction and that their results have been sufficiently comprehensive and credible for them to be taken seriously by policy makers. The climate models which have been developed are probably the most elaborate and sophisticated of computer models developed in any area of natural science.

A great deal remains to be done to narrow the uncertainty of model predictions. The first priorities must be to improve the modelling of clouds and the description in the models of the ocean-atmosphere interaction. Larger and faster computers are required to tackle this problem, especially to enable the resolution of the model grid to be increased, as well as more sophisticated model physics and dynamics. Much more thorough observations of all components of the climate system are also necessary, so that more accurate validation of the model formulations can be achieved. Very substantial national and international programmes are underway to address these issues.

Questions

1   Make an estimate of the speed in operations per second of Richardson's 'people' computer. Where does it fall on Fig. 5.1?

2   If the spacing between the grid points in a model is 100 km and there are 20 levels in the vertical, what is the total number of grid points in a global model? If the distance between grid points in the horizontal is halved, how much longer will a given forecast take to run on the computer?

3   Take your local weather forecasts over a week and describe their accuracy for 12, 24 and 48 hours ahead.

4   Estimate the average energy received from the sun over a square region of the ocean surface, one side of the square being a line between northern Europe and Iceland. Compare with the average transport of energy into the region by the north Atlantic ocean (Fig. 5.16).

5   Take a hypothetical situation in which a completely absorbing planetary surface at a temperature of 280 K is covered by a non-absorbing and non-emitting atmosphere. If a cloud which is non-absorbing in the visible part of the spectrum but completely absorbing in the thermal infrared is present above the surface, show that its equilibrium temperature will be 235 K ($=280/2^{0.25}$ K)[34]. Show also that if the cloud reflects 50% of solar radiation, the rest being transmitted, the planet's surface will receive the same amount of energy as when the cloud is absent. Can you substantiate the statement that the presence of low clouds tends to cool the Earth while high clouds tend towards warming of it?

6   Associated with the melting of sea-ice which results in increased evaporation from the water surface, additional low cloud can appear. How does this affect the ice-albedo feedback? Does it tend to make it more or less positive?

7   It is sometimes argued that weather and climate models are the most sophisticated and soundly based models in natural science. Compare them (e.g. in their assumptions, their scientific basis, their potential accuracy etc.) with other computer models with which you are familiar both in natural science and social science (eg models of the economy).

Notes

1 Further information regarding the subject of this chapter can be found in: J. T. Houghton, *The Physics of Atmospheres*, CUP, second edition, 1986; *Climate Change, the IPCC Scientific Assessment*, eds. J. T. Houghton, G. J. Jenkins and J. J. Ephraums, CUP, 1990; *Climate Change 1992, the Supplementary Report to the IPCC Scientific Assessment*, eds. J. T. Houghton, B. A. Callander and S. K. Varney, CUP, 1992; *Climate Change 1995: The Science of Climate Change*, eds. J. T. Houghton, L. G. Meira Filho, B. A. Callender, N. Harris, A. Kattenberg and K. Maskell, CUP, 1996; *Climate System Modelling*, ed. K. E. Trenberth, CUP, 1992; J. T. Houghton, 'The Bakerian Lecture, 1991: The predictability of weather and climate', *Phil. Trans. R. Soc. Lond. A* (1991), **337**, pp. 521–72.

2 L. F. Richardson, *Weather Prediction by Numerical Processes*, CUP, 1922, reprinted Dover, 1965.

3 For more detail see for instance J. T. Houghton, *The Physics of Atmosphere*, CUP, second edition, 1986.

4 After S. Milton, Meteorological Office, quoted by J. T. Houghton, 'The Bakerian Lecture, 1991: The predictability of weather and climate', *Phil. Trans. R. Soc. Lond. A* (1991), **337**, pp. 521–72.

5 For more detail see J. T. Houghton, 'The Bakerian Lecture, 1991: The predictability of weather and climate', *Phil. Trans. R. Soc. Lond. A* (1991), **337**, pp. 521–72; also T. N. Palmer, 'A nonlinear perspective on climate change', *Weather*, **48**, 1993, pp. 314–26.

6 An equation such as $y = ax + b$ is linear; a plot of y against x is a straight line. Examples of non-linear equations are $y = ax^2 + b$ or $y + xy = ax + b$; plots of y against x for these equations would not be straight lines. In the case of the pendulum, the equations describing the motion are only approximately linear for very small angles from the vertical where the sine of the angle is approximately equal to the angle; at larger angles this approximation becomes much less accurate and the equations are non–linear.

7 After J. Lighthill, 'The recently recognized failure in Newtonian dynamics', *Proc. Roy. Soc. Lond. A* **407**, 1986, pp. 35–50.

8 *The storm 15/16 October 1987*, UK Meteorological Office Report, Bracknell.

9 For instance, J. Shukla and M. J. Fennessy, *J. Atmos. Sci.*, **45**, 1988, pp. 9–28.

10 See for instance M. A. Cane in *Climate System Modelling*, ed. K. E. Trenberth, CUP 1992, chapter 18, pp. 583–616.

11 Data from S. E. Nicholson and C. K. Folland, see J. T. Houghton, 'The Bakerian Lecture, 1991: The predictability of weather and climate', *Phil. Trans. R. Soc. Lond. A* (1991), **337**, pp. 521–72.

12 C. K. Folland *et al.*, 'Sahel rainfall and worldwide sea temperature 1901–1985', *Nature*, **320**, 1986, pp. 602–7.

13 M. Hulme *et al.*, 'Seasonal rainfall forecasting for Africa: Part 1, Current status and future developments', *Int. J. Environ. Stud. A* **39**, 1992, pp. 245–56.

14 From J. T. Houghton, 'The Bakerian Lecture, 1991: The predictability of weather and climate', *Phil. Trans. R. Soc. Lond. A* (1991), **337**, pp. 521–72.

15 Associated with water vapour feedback is also *lapse rate feedback* which occurs because associated with changes of temperature and water vapour content in the troposphere are changes in the average lapse rate (the rate of fall of temperature with height). Such changes lead to this further feedback which is generally much smaller in magnitude than water vapour feedback but of the opposite sign i.e. negative instead of positive. Frequently, when overall values for water vapour feedback are quoted the lapse rate feedback has been included.

16 Diagram from Catherine Senior, UK Meteorological Office.

17 From J. D. Woods, 'The upper ocean and air sea interaction in global climate' in *The Global Climate*, ed. J. T. Houghton, CUP, 1984, pp. 141–87.

18 R. S. Lindzen in a paper 'Some coolness concerning global warming', *Bull. Amer. Met. Soc.*, **71**, 1990, pp. 288–99, has queried the magnitude and sign of the feedback due to water vapour especially in the upper troposphere and has suggested that it could be much less positive than predicted by models and could even be slightly negative. His arguments have been reviewed by W. L. Gates *et al.*, 'Climate modelling, climate prediction and model validation' in *Climate Change 1992: the Supplementary Report to the IPCC Assessment*, eds. J. T. Houghton, B. A. Callander and S. K. Varney, CUP, 1992, pp. 97–134, and by R. E. Dickinson *et al.*, 'Climate processes', chapter 4 pp. 193–227, in *Climate Change 1995: The Science of Climate Change*, eds. J. T. Houghton, L. G. Meira Filho, B. A. Callender, N. Harris, A. Kattenberg and K. Maskell, CUP, 1996. While some uncertainty remains about the role of upper tropospheric water vapour, the weight of evidence from both observations and models is that the overall water vapour feedback is positive and of about the magnitude which is simulated by the models.

19 J. F. B. Mitchell, C. A. Senior and W. J. Ingram, *Nature*, **341**, 1989, pp. 132–4.

20 C. A. Senior and J. F. B. Mitchell 'Carbon dioxide and climate: the impact of cloud parameterization', *J. Climate*, **6**, 1993, pp. 393–418.

21 See *Climate Change, the IPCC Scientific Assessment*, eds. J. T. Houghton, G. J. Jenkins and J. J. Ephraums, CUP, 1990, and *Climate Change 1995: The Science of Climate Change*, eds. J. T. Houghton, L. G. Meira Filho, B. A. Callender, N. Harris, A. Kattenberg and K. Maskell, CUP, 1996.

22 W. S. Broecker and G. H. Denton, 'What drives glacial cycles?', *Scien. Amer.* **262**, 1990, pp. 43–50.

23 S. Manabe and R. J. Stouffer, 'Century-scale effects of increased atmospheric CO_2 on the ocean–atmosphere system', *Nature*, **364**, 1993, pp. 215–18; details also in *Climate Change, the IPCC Scientific Assessment*, eds. J. T. Houghton, G. J. Jenkins and J. J. Ephraums, CUP, 1990, chapter 6; and *Climate Change 1995: The Science of Climate Change*, eds. J. T. Houghton, L. G. Meira Filho, B. A. Callender, N. Harris, A. Kattenberg and K. Maskell, CUP, 1996, chapter 4.

24 After W. S. Broecker and G. H. Denton, 'What drives glacial cycles?', *Sci. Amer.* **262**, 1990, pp. 43–50.

25 For more detail see W. L. Gates *et al.*, 'Climate Models – Evaluation', chapter 5 in *Climate Change 1995: The Science of Climate Change*, eds. J. T. Houghton, L. G. Meira Filho, B. A. Callender, N. Harris, A. Kattenberg and K. Maskell, CUP, 1996.

26 This diagram and information about modelling past climates is from J. E. Kutzbach in *Climate System Modelling*, ed. K.E. Trenberth, CUP, 1992, pp. 669–701.

27 J. L. Sarmiento, *J. Phys. Oceanog.*, **13**, 1983, pp. 1924–39.

28 J. Hansen *et al.*, Potential climate impact of Mt Pinatubo eruption, *Geophys. Res. Lett.* **19**, 1992, pp. 215–18; also quoted in *Climate Change 1995: The Science of Climate Change*, eds. J. T. Houghton, L. G. Meira Filho, B. A. Callender, N. Harris, A. Kattenberg and K. Maskell, CUP, 1996, Technical Summary.

29 H.-F. Graf *et al.*, 'Pinatubo eruption winter climate effects: model versus observations,' *Climate Dynamics*, **9**, 1993, pp. 61–73.

30 Data from model at the Hadley Centre, UK Meteorological Office – from *Climate Change 1995: The Science of Climate Change*, eds. J. T. Houghton, L. G. Meira Filho, B. A. Callender, N. Harris, A. Kattenberg and K. Maskell, CUP, 1996, Technical Summary.

31 *Climate Change, the IPCC Scientific Assessment*, eds. J. T. Houghton, G. J. Jenkins and J. J. Ephraums, CUP, 1990, Policymakers' Summary.

32 *Climate Change 1995: The Science of Climate Change*, eds. J. T. Houghton, L. G. Meira Filho, B. A. Callender, N. Harris, A. Kattenberg and K. Maskell, CUP, 1996, Summary for Policymakers.

33 From K. E. Taylor and J. E. Penner, *Nature* **369**, 1994, pp. 734–6, also quoted in B. D. Santer *et al.*, 'Detection of Climate Change and Attribution of Causes', chapter 8 in *Climate Change 1995: The Science of Climate Change*, eds. J. T. Houghton, L. G. Meira Filho, B. A. Callender, N. Harris, A. Kattenberg and K. Maskell, CUP, 1996.

34 Hint: recall Stefan's blackbody radiation law that the energy emitted is proportional to the fourth power of the temperature

6 Climate Change under Business-as-usual

The last chapter showed that the most effective tool we possess for the prediction of future climate change due to human activities is the climate model. This chapter will describe the predictions of models for likely climate change next century. It will also consider other factors which might lead to climate change and assess their importance relative to the effect of greenhouse gases.

Model projections

A principal reason for the development of climate models is to learn about some of the detail of the likely climate change next century and beyond. Because model simulations into the future depend on assumptions regarding the future anthropogenic emissions of greenhouse gases, which in turn depend on assumptions about many factors involving human behaviour, it has been thought inappropriate and possibly misleading to call the simulations of future climate so far into the future 'predictions'. They are therefore generally called 'projections' to emphasize that what is being done is to explore the likely future climates which will arise from a range of assumptions regarding human activities.

Results which come from the most sophisticated coupled atmosphere-ocean models of the kind described in the last chapter provide fundamental information on which to base climate projections. However, because they are so demanding on computer time only a limited number of results from such models are available. Many studies have also therefore been carried out with simpler models. Some of these, while possessing a full description of atmospheric processes, only have a simplified description of the ocean; these can be useful in exploring regional change, Others, sometimes called energy balance models (see box), drastically simplify the dynamics and physics of both atmosphere and ocean and are useful in exploring changes in the global average response with widely different emission scenarios. Results from simplified models need to be carefully compared with those from the best coupled atmosphere-ocean models and the simplified models 'tuned' so that, for the particular parameters for which they are being employed, agreement with the more complete models is as close as possible. The projections presented in the next sections depend on results from all these kinds of models.

In order to assist comparison between models, experiments with many models have been run with the atmospheric concentration of carbon dioxide doubled from its pre-industrial level of 280 ppmv. The global average temperature rise under steady conditions of doubled carbon dioxide concentration has

Simple climate models

In Chapter 5 a detailed description was given of general circulation models (GCMs) of the atmosphere and the ocean and of the way in which they are coupled together (in AOGCMs) to provide simulations of the current climate and of climate perturbed by anthropogenic emissions of greenhouse gases. These models provide the basis of our projections of the detail of future climate. However, because they are so elaborate, they take a great deal of computer time so that only a few simulations can be run with these large coupled models.

To carry out more simulations under different future emission profiles of greenhouse gases or of aerosols or to explore the sensitivity of future change to different parameters (for instance parameters describing the feedbacks in the atmosphere which largely define the climate sensitivity), extensive use has been made of simple climate models[1]. These simpler models are 'tuned' so as to agree closely with the results of the more complex AOGCMs in cases where they can be compared. The most radical simplification in the simpler models is to remove one or more of the dimensions so that the quantities of interest are averaged over latitude circles (in two-dimensional models) or over the whole globe (in one-dimensional models). Such models can, of course, only simulate latitudinal or global averages – they can provide no regional information.

Figure 6.1 illustrates the components of such a model in which the atmosphere is contained within a 'box' with appropriate radiative inputs and outputs. Exchange of heat occurs at the land surface (another 'box') and the ocean surface. Within the ocean allowance is made for vertical diffusion and vertical circulation. Such a model is appropriate for simulating changes in global average surface temperature with increasing greenhouse gases or aerosols. When exchanges of carbon dioxide across the interfaces between the atmosphere, the land and the ocean are also included, the model can be employed to simulate the carbon cycle.

Fig. 6.1
The components of a simple 'upwelling-diffusion' climate model.

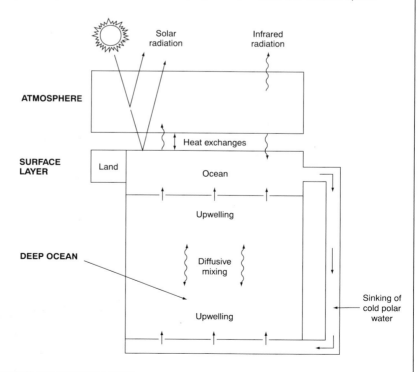

become known as the climate sensitivity[2]. The Intergovernmental Panel on Climate Change (IPCC) in its 1990 report gave a 'best estimate' of 2.5 °C for the climate sensitivity; it also considered that it was unlikely to lie outside the range of 1.5–4.5 °C. The IPCC 1995 Report has confirmed these values. The projections presented in this chapter follow the IPCC Assessments[3].

To make projections of climate change, knowledge of future changes in greenhouse gas concentrations is first required; this has already been addressed in Chapter 3. The projections presented in this chapter have used the greenhouse gas emissions scenario IS 92a, published in the IPCC 1992 report[4] (see Fig. 3.5 and Table 3.2) which, in the absence of any strong controls on emissions is the one based on the most likely assumptions about future conditions. For this reason I have called it the 'business-as-usual' emissions scenario. The method by which the emissions scenario is turned into future projections of greenhouse gas concentrations and into radiative forcing (Fig. 3.9) has already been described in Chapter 3.

Projections of global average temperature

When radiative forcing information of the kind presented in Fig. 3.9 is incorporated into the models, projections of climate change can be made. A useful proxy for climate change which has been widely used is the change in global average temperature.

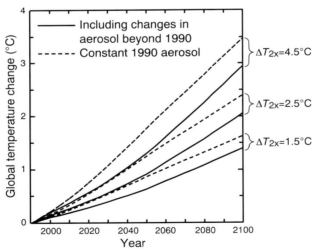

Fig 6.2
The projected change in global average surface air temperature under a 'business-as-usual' scenario (IPCC IS 92a) from 1990 until 2100[5]. To obtain the rise from pre-industrial times (allowing for both the increase in greenhouse gases and the effects of aerosol to date), add about 0.5 °C. Curves are also shown for the assumption that the aerosol concentrations remain constant at 1990 values. The middle curves are the IPCC's best estimate; the upper and lower curves indicate the estimated range of uncertainty. The three curves correspond to 'climate sensitivities' of 4.5, 2.5 and 1.5 °C respectively.

The projected rise in global average temperature due to the increase in greenhouse gases and aerosols from 1990 to 2100 is shown in Fig. 6.2, from which it will be seen that the best estimate of the rise in global average temperature from now until the end of next century is about 2 °C (about 2.5 °C from preindustrial times), with an uncertainty range from about 1.5 to 3.5 °C. Compared with the temperature changes normally experienced from day to day and throughout the year, 2 °C does not seem very much. But, as was pointed out in Chapter 1, it is in fact a large amount when considering a globally averaged temperature. Compare the 5 or 6 °C change in global average temperature which occurs between the middle of an ice age and the warm period in between ice ages; 2 °C is about a third of an ice age in terms of the amount of climate change!

Figure 6.2 shows two sets of curves, one assuming that aerosol concentrations grow as estimated by the IS 92a scenario (Table 3.2) and the other assuming that they stay constant at 1990 levels. There are several reasons for presenting the information both with and without the aerosol contributions. Firstly, as was explained in Chapter 3, the likely future concentrations of aerosols are particularly uncertain because of other environmental problems such as acid rain. It has been argued that, because of action which will be taken to mollify the acid rain problem, the IS 92a scenario, so far as it projects sulphur emissions, is likely to be too high[6]. Secondly, when considering global average temperature and its impact on, for instance, sea level rise (see Chapter 7) it is appropriate to use the projections which include the effects of increasing aerosol. However, because the effects of aerosol forcing are far from uniform over the globe (Fig. 3.7), it is not correct, when considering climate change and its regional characteristics, to consider the effects of increasing aerosol as a simple offset to those of the increase in greenhouse gases. Detailed regional information from the best climate models needs to be employed to assess the climate change under different assumptions about the increases in both greenhouse gases and aerosols. Reference to this will be made again in the next section.

By about the year 2030 when the equivalent amount of carbon dioxide in the atmosphere (see Chapter 3) compared with pre-industrial times will have doubled, the best estimate of the temperature rise to be expected from the increase in greenhouse gases (ignoring for the moment the effect of aerosols) is about 0.8 °C from now and about 1.6 °C from pre-industrial times[7]. As Chapter 5 showed, this is less than the 2.5 °C which would be expected for doubled carbon dioxide under steady conditions because of the slowing effect of the oceans on the temperature rise. But this means that, as the carbon dioxide concentration continues to increase, at any given time there exists a commitment to further significant temperature rise which has not been realised at that time.

The rate of change of global average temperature projected for next century is in the range of 0.15–0.3 °C per decade with a best estimate of about 0.2 °C per decade. Again this seems a small amount; most people would find it hard to detect a change in temperature of a fraction of a degree. But remembering that these are global averages, these rates of change become very large. Indeed, they are much larger than rates of change for the past ten thousand years inferred from paleoclimate data (Fig. 6.3). As we shall see in the next chapter,

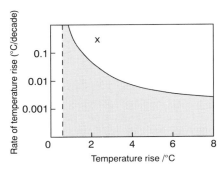

Fig. 6.3
Within the shaded region are typical rates of climate change during the past 10,000 years, over periods up to about a century, plotted against actual change deduced from climate records and paleoclimate data[8]. Also plotted (X) is the projected rate of change next century due to a business-as-usual increase of greenhouse gases – well outside the shaded region.

the ability of ecosystems to adapt to climate change depends critically on the rate of change. For many ecosystems 0.2 °C per decade would be a very rapid rate of change indeed.

Regional climate change

Discussing changes in global average temperature can produce some kind of overall idea of the magnitude of likely climate change, but in terms of the regional implications, a global average conveys rather little information. What is required is spatial detail. It is in the regional or local changes that the effects and impacts of global climate change will be felt.

An important point to realise concerning regional climate change is that, because of the way the atmospheric circulation operates and the interactions which govern the behaviour of the whole climate system, it should not be expected that climate change over the globe should be at all uniform. We can for instance expect substantial differences between the changes over large land masses and over the ocean; land possesses a much smaller thermal capacity and so can respond more quickly. The response can also be expected to vary with latitude.

Further, much natural climate variability occurs because of changes in or oscillations between persistent climatic patterns or regimes. The Pacific, North Atlantic Anomaly (PNA) (which is dominated by high pressure over the eastern Pacific and western North America and which tends to lead to very cold winters in the eastern United States) and the El Niño events mentioned in Chapter 5 are examples of such regimes. Important components of climate change in response to the forcing due to the increase in greenhouse gases might be expected to be in the form of changes in the intensity or the frequency of established climate patterns illustrated by these regimes[9].

For instance, it has been suggested that the very unusual pattern of El Niño events since 1990 might be linked to the increase in greenhouse gases during recent decades. Because of the large influence of El Niño events on the climate (especially on the incidence of extreme climate events) in many parts of the world, it is especially important to establish the connections that might exist between anthropogenic forcing of the climate and the nature of the climate

Fig. 6.4
The pattern of annual surface air temperature change in °C projected at the time of doubling of atmospheric carbon dioxide concentration by a transient coupled atmosphere ocean model, for increase in carbon dioxide only (a) and increase in carbon dioxide and aerosols (b)[10].

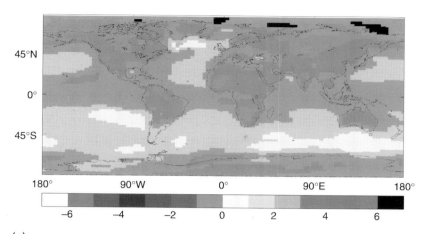

(a)

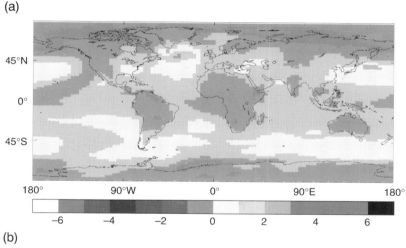

(b)

response as it might affect the character of quasi-regular events such as the El Niño.

A major difficulty with providing good information on the regional scale is that of the inadequate information we have regarding the likely future distribution of increased sulphate aerosols, the effect of which on the regional scale can be large. However there are some broad features which are shown by all model simulations whether they are forced with increased greenhouse gases alone or with both increased greenhouse gases and aerosols (Fig 6.4). These are[11]:

- generally greater surface warming of the land than of the oceans in winter;
- minimum warming around Antarctica and in the northern North Atlantic which is associated with deep oceanic mixing in those areas;
- maximum warming in high northern latitudes in late autumn and winter associated with reduced sea ice and snow cover;
- little warming over the Arctic in summer;
- little seasonal variation of the warming in low latitudes or over the southern circumpolar ocean;

- a reduction in diurnal temperature range over land in most seasons and most regions;
- an enhanced global mean hydrological cycle;
- increased precipitation in high latitudes in winter.

Including the effects of increased aerosols (based on the IS 92a scenario of sulphur emissions) in simulations of future climate leads to a somewhat reduced surface warming, mainly in mid-latitudes of the northern hemisphere; also the maximum winter warming at high northern latitudes is less extensive. Continental patterns of change can also be affected by the global pattern of increases in aerosols. For example when increases in both greenhouse gases and aerosols are included (with a large increase in aerosols over Asia) Asian summer monsoon rainfall decreases, whereas in simulations with only the effect of greenhouse gases represented, rainfall in that area increases (see box in Chapter 7 on page 123). Over southern Europe the opposite occurs; the influence of aerosols in addition to greenhouse gases leads to increased summer rainfall whereas it decreases in simulations with greenhouse gases only. These examples are clearly sensitive to the particular aerosol scenario used; different aerosol amounts and patterns are likely to give different results. More model simulations with different aerosol amounts and distributions need to be carried out.

Soil moisture is an important parameter for agriculture; in regions with marginal rainfall it is critical. Changes in soil moisture depend on changes in precipitation and also on changes in surface air temperature leading to increased evaporation. Models in which increases in greenhouse gases only are included consistently show decreased soil moisture in summer in mid-latitude regions such as the continental United States, southern Europe and parts of Australia. The inclusion of patterns of increased aerosols (as in the IS 92a scenario) in these simulations can significantly change these results.

So much for change on the continental scale. Can more specific information be provided about change for smaller regions? We have already seen the difficulty which arises from our lack of knowledge of future aerosol emissions. A further problem is the limited capability of climate models on the smaller scale due particularly to two factors. Firstly, the model grid-size in the horizontal is coarse, typically 300 km or more; topographic and other features on a smaller scale than the grid-size cannot be included with the result that the description of local and regional climate is severely affected. Secondly, as Chapter 5 explains, models deal inadequately with clouds, a problem which particularly influences regional detail. A great deal of research is going on to overcome these problems and to improve the capability of models to provide regional predictions. One promising technique is the use of 'nested' models; a high resolution model for a given region is 'nested' in a much coarser model for the globe. However, it is important to realize that, even if the models were perfect, because much more natural variability is apparent in local climate than in climate averaged over continental or larger scales, projections on the local and regional scale are bound to be more uncertain than those on larger scales.

Changes in climate extremes

The last section looked at the likely regional patterns of climate change. Can anything be said about likely changes in the frequency or intensity of climate extremes in the future? It is, after all, not the changes in average climate which are noticeable, but the extremes of climate – the droughts, the floods, the storms and the extremes of temperature in very cold or very warm periods – which provide the largest impact on our lives, as discussed in Chapter 1.

The most obvious change we can expect in extremes is a large increase in the number of extremely warm days (Fig. 6.5) coupled with a decrease in the number of extremely cold days. However, the changes which are likely to give most impact are those connected with the hydrological cycle. Although, for the reasons we have explained, the models do not yet provide much reliable information about the distribution of likely changes in precipitation, they all agree that in the warmer world with increased greenhouse gases the hydrological cycle will, on average, become more intense. This means, as we have already noted, that the average precipitation will tend to increase especially at high latitudes. It also means that there will tend to be more periods of heavy rainfall. Perhaps surprisingly, there will also tend to be regions where the precipitation is reduced. This is because, in regions of convection the areas of downdraught become drier as the areas of updraught become more moist. What effect are these changes likely to have on the incidence of droughts and floods – these we noted in Chapter 1 are the climate extremes which cause the greatest problems?

Fig 6.5
Illustrating that, when the average temperature increases, there is likely to be a large increase in the number of extremely warm days[12]. In the diagram a curve describing the probability distribution of days with given temperatures is shifted to represent a warmer overall average.

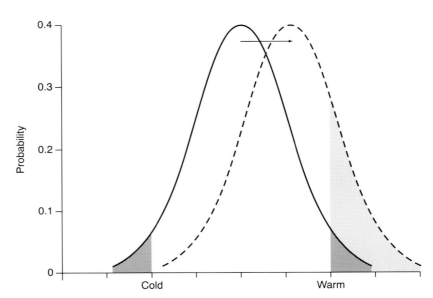

Fig. 6.6
Daily rainfall for Italy for a three-year period simulated using a climate model: (a) for the current climate situation and (b) for the climate if the equivalent carbon dioxide concentration increased by a factor of 4 from its pre-industrial value (predicted to occur under the IS 92a scenario by about 2100)[13].

Take, for instance, the likelihood of drought in regions where the average summer rainfall falls by 10 to 20 per cent as is likely to occur, for instance, in southern Europe under a scenario with increased greenhouse gases but with little increase in sulphate aerosol. The likely result of such a drop in rainfall is not that the number of rainy days will remain the same, with less rain falling each time. It is more likely that there will be substantially fewer rainy days and considerably more chance of prolonged periods of no rainfall at all (see Fig. 6.6); in other words, much more likelihood of drought. Further, the higher temperatures will lead to increased evaporation reducing the amount of moisture available at the surface – thus adding to the drought conditions. The proportional increase in the likelihood of drought is greater than the proportional decrease in average rainfall.

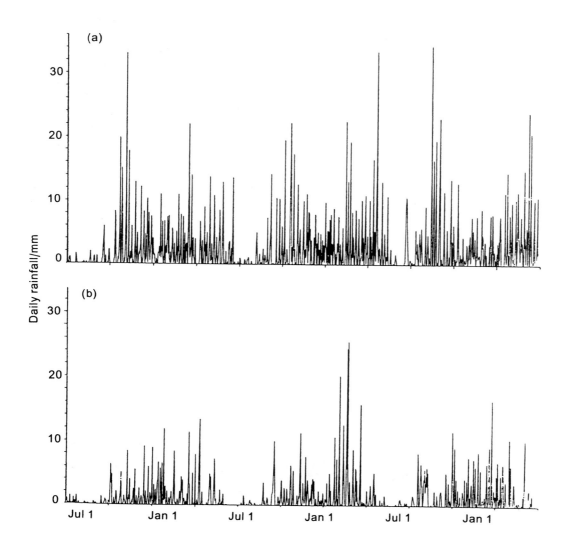

A similar story emerges when considering what might occur in regions of increased rainfall. Often in such regions the larger amounts of rainfall will come from increased convective activity: more really heavy showers and more intense thunderstorms. The result of a study with an Australian climate model of the effect on rainfall amounts of doubling the carbon dioxide concentration is shown in Fig. 6.7. The total rainfall was not much changed, but the number of days with small amounts of rain decreased while the number of days with large rainfall (greater than 25 mm) doubled. The study suggests that the probability of conditions leading to floods would at least have doubled. Similar results (fewer rainy days, higher maximum daily rainfalls for a given mean rainfall rate) have been obtained from the results of other climate models.

Thus in the warmer world of increased greenhouse gases, different places will experience more frequent droughts and floods. What about other climate extremes, intense storms, for instance? How about hurricanes and typhoons, the violent rotating cyclones which are found over the tropical oceans and which cause such devastation when they hit land? The energy for such storms largely comes from the latent heat of the water which has been evaporated from the warm ocean surface and which condenses in the clouds within the storm, releasing energy. It might be expected that warmer sea temperatures would mean more energy release, leading to more frequent and intense storms. However, ocean temperature is not the only parameter controlling the genesis of tropical storms; the nature of the overall atmospheric flow is also important. Models can take all these factors into account but, because of the relatively large size of their grid, they are unable to simulate very well the detail of relatively small disturbances like tropical cyclones. Because simulations of tropical cyclones from different models have produced quite different results, model projections of the influence of increased greenhouse gas concentrations on the frequency and intensity of tropical cyclones must therefore, as yet, be treated with some caution.

Fig. 6.7
Changes in the frequency of the occurrence of different daily rainfall amounts with doubled carbon dioxide as estimated by the CSIRO climate model in Australia[14].

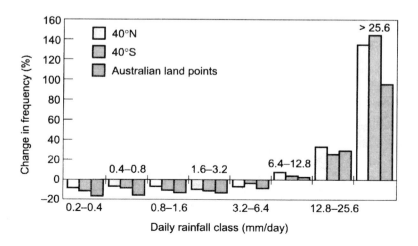

The various factors which control the incidence of storms at mid-latitudes are complex. Two factors tend to an increased intensity of storms. The first, as with tropical storms, is that higher temperatures, especially of the ocean surface, tend to lead to more energy being available. The second factor is that the larger temperature contrast between land and sea, especially in the northern hemisphere, tends to generate steeper temperature gradients, which in turn generate stronger flow and greater likelihood of instability. The region around the Atlantic seaboard of Europe is one area where such increased storminess might be expected, a result which some model simulations have shown. However, such a picture may well be too simple; other models show changes in storm tracks which result in very different changes in some regions. Much careful study with climate models is going on, aimed at the improvement of understanding and confidence in the detailed predictions of future change, especially those concerned with extremes.

Longer-term climate change

Most of the projections of future climate which have been published cover the next century. For instance, the curves plotted in Fig. 6.2 go up to the year 2100. They illustrate what is likely to occur if fossil fuels continue to provide most of the world's energy needs during that period.

From the beginning of the industrial revolution until 1990 the burning of fossil fuels released approximately 250 Gt of carbon in the form of carbon dioxide into the atmosphere. Under the IS 92a scenario it is projected that a further 1,400 Gt will be released by the year 2100. As Chapter 11 will show, the reserves of fossil fuels in total are sufficient to enable their rate of use to continue to grow well beyond the year 2100. If that were to happen the global average temperature would continue to rise and could, in the twenty-second century, reach very high levels, perhaps between 5 and 10 °C higher than today (see Chapter 9). The associated changes in climate would be correspondingly large and could well be irreversible.

Especially when considering the longer term, there is also the possibility of surprises – changes in the climate system that are unexpected. The discovery of the 'ozone hole' is an example of a change in the atmosphere due to human activities, which was a scientific 'surprise'. By their very nature such 'surprises' cannot, of course, be foreseen. However, there are various parts of the system which are, as yet, not well understood, where such possibilities might be looked for, for instance, in the deep ocean circulation (see box in Chapter 5, p. 81) or in feedbacks in the biosphere (see box in Chapter 3, p. 31).

Other factors which might influence climate change

So far climate change due to human activities has been considered. Are there other factors, external to the climate system, which might induce change? Chapter 4 showed that it was variations in the incoming solar energy as a result of changes in the Earth's orbit which triggered the ice ages and the major

Does the sun's output change?

Some scientists have suggested that climate variations, even short-term ones, might be the result of changes in the sun's energy output. Such suggestions are bound to be somewhat speculative because no actual measurements of such changes exist. Accurate measurements of solar output have only been available since 1978, from satellites outside the disturbing effects of the Earth's atmosphere. These measurements indicate a very constant solar output, changing by about 0.1 per cent between a maximum and a minimum in the cycle of solar magnetic activity indicated by the number of sunspots.

It is known from astronomical records and from measurements of radioactive carbon in the atmosphere that this solar sunspot activity has, from time to time over the past few thousand years, shown large variations. Of particular interest is the period known as the Maunder Minimum in the 17th century when very few sunspots were recorded[15]. Studies of the recent measurements of solar output correlated with other indicators of solar activity, when extrapolated to this earlier period, suggest that the sun was a little less bright in the 17th century, perhaps by about 0.4 per cent or about 1 watt per square metre in the solar irradiance (the average solar energy incident on the Earth's surface). This reduction in solar energy may have been a cause of the cooler period at that time known as the 'Little Ice Age'. Careful studies have estimated that the maximum variations in solar irradiance since 1850 are unlikely to be greater than about 0.5 watts per square metre. This is about the same as the change in the energy regime at the Earth's surface due to about ten years' increase in greenhouse gases at the current rate.

climate changes of the past. These variations are, of course, still going on; what influence are they having now?

Over the past 10,000 years, because of these orbital changes, the solar radiation incident at 60 °N in July has decreased by about 35 watts per square metre, which is quite a large amount. But over 100 years the change is only at most a few tenths of a watt per square metre, which is much less than the changes due to the increases in greenhouse gases (remember that doubling carbon dioxide alters the thermal radiation by about 4 watts per square metre – see Chapter 2).

These orbital changes only alter the *distribution* of incoming solar energy over the Earth's surface; the total amount of energy reaching the Earth is hardly affected by them. Suggestions have been made that the actual energy output of the sun might change with time. As we mentioned in Chapter 3 (see Fig. 3.8) and as is described in the box, such changes if they occur, are estimated to be much smaller than changes in the energy regime at the Earth's surface due to the increase in greenhouse gases.

Another influence on climate comes from volcanic eruptions. Their effects, lasting typically a few years, are relatively short-term compared with the much longer-term effects of the increase of greenhouse gases. The recent large volcanic eruption of Mount Pinatubo in the Philippines which occurred in June 1991 has already been mentioned. Estimates of the change in the net amount of radiation (solar and thermal) at the top of the atmosphere resulting from this eruption are of about 0.5 watts per square metre. This perturbation lasts for about two or three years while the major part of the dust settles out of the atmosphere; the longer-term change in radiation forcing, due to the minute particles of dust which last in the stratosphere for somewhat longer, is much smaller.

To summarize this chapter:

- The increase in greenhouse gases is by far the largest of the factors which can lead to climate change during next century.
- The likely climate change for a business-as-usual scenario of greenhouse gas emissions has been described in terms of global average temperature and in terms of regional change of temperature and precipitation and the occurrence of extremes.
- The rate of change is likely to be larger than the Earth has seen at any time during the past 10,000 years.
- The changes which are likely to have the greatest impact will be changes in the frequencies, intensities and locations of climate extremes, especially droughts and floods.
- Sufficient fossil fuel reserves are available to provide for continuing growth in fossil fuel emissions of carbon dioxide well into the twenty-second century. If this occurred the climate change could be very large indeed and have unpredictable features or 'surprises'.

The next chapter will look at the impact of such changes on sea level, on water, on food supplies and on human health. Later chapters of the book will then suggest what action might be taken to slow down and eventually to terminate the rate of change.

Questions   1   Suggest, for Fig. 6.5, an appropriate temperature scale for a place you know. Define what might be meant by a very hot day and estimate the percentage increase in the probability of such days if the average temperature increases by 1, 2 and 4 °C.

  2   From Fig. 6.6, compare the maximum length of periods in the summer with less than 1 mm, 2 mm, 5 mm of total rainfall under normal climate conditions and under conditions with increased carbon dioxide.

  3   It is sometimes suggested that IPCC estimates of global warming have been continually reduced with successive IPCC reports since 1970, so that the problem is now less important than it was. How far do you agree with this statement?

  4   Look at the assumptions underlying the full range of IPCC emission scenarios in the IPCC 1992 Report. How much do you agree that IS 92a is the most likely scenario?

5 It is sometimes suggested that northwest Europe could become colder in the future while most of the rest of the world becomes warmer. What could cause this and how likely do you think it is to occur?

6 How important do you consider it is to emphasise the possibility of 'surprises' when presenting projections of likely future climate change.

Notes

1 See *IPCC Technical Paper 2*, L. D. D. Harvey *et al.*,'An introduction to simple climate models used in the IPCC Second Assessment Report', IPCC, Geneva, 1997.

2 Because the response of global average temperature to the increase of carbon dioxide is logarithmic in the carbon dioxide concentration, the increase of global average temperature for doubling of carbon dioxide concentration is the same whatever the concentration which forms the base for the doubling, eg doubling from 280 ppmv or from 360 ppmv produces the same rise in global average temperature.

3 *Climate Change, the IPCC Scientific Assessment*, eds. J. T. Houghton, G. J. Jenkins and J. J. Ephraums, CUP, 1990; *Climate Change 1995: The Science of Climate Change*, eds. J. T. Houghton, L. G. Meira Filho, B. A. Callender, N. Harris, A. Kattenberg and K. Maskell, CUP, 1996.

4 J. Leggett, W. J. Pepper and R. J. Swart, 'Emissions scenarios for the IPCC: an update' in *Climate Change 1992: the Supplementary Report to the IPCC Assessment*, eds. J. T. Houghton, B. A. Callander and S. K. Varney, CUP, 1992, pp. 69–95. Small modifications have been made to the IS 92a scenario to take into account developments in the Montreal Protocol; these are described in the caption to Fig. 3.9.

5 From A. Kattenberg *et al.*, 'Climate models – projections of future climate', chapter 6 in *Climate Change 1995: The Science of Climate Change*, eds. J. T. Houghton, L. G. Meira Filho, B. A. Callender, N. Harris, A. Kattenberg and K. Maskell, CUP, 1996.

6 The World Energy Council report *Energy for tomorrow's world*, 1993, (World Energy Council, London) in its most likely scenario projects that global sulphur emissions in 2020 will be almost the same as in 1990, although with a different distribution (more over Asia but less over Europe and North America). An extension to this study (*Global Energy Perspectives to 2050 and beyond*: World Energy Council Report 1995) projects at 2050 global sulphur emissions which are little more than half of 1990 levels.

7 Note that 0.8 ºC has been added for the temperature rise from pre–industrial times to the present arising from the increase in greenhouse gases only. When aerosol effects are included the rise is reduced to about 0.5 ºC.

8 Adapted from Climate *Change: Meeting the Challenge*, Report by a Commonwealth Group of Experts, Commonwealth Secretariat, London, 1989, p. 33; (original source of diagram: Sassin *et al.*, 1988).

9 For more on this see T. N. Palmer, *Weather*, **48**, 1993, pp. 314–25

10 From J. F. B. Mitchell *et al.*, *J. Climate*, **10**, 1995, pp. 2364–86; also presented in *Climate Change 1995: The Science of Climate Change*, eds. J. T. Houghton, L. G. Meira Filho, B. A. Callender, N. Harris, A. Kattenberg and K. Maskell, CUP, 1996, Technical Summary.

11 From *Climate Change 1995: The Science of Climate Change*, eds. J. T. Houghton, L. G. Meira Filho, B. A. Callender, N. Harris, A. Kattenberg and K. Maskell, CUP, 1996, Technical Summary.

12 Adapted from K. Maskell, I. M. Mintzer and B. A. Callender, 'Basic science of climate change', *Lancet*, **342**, 1993, pp. 1027–31.

13 From C. A. Wilson and J. F. B. Mitchell, 'Simulation of climate and CO_2 induced climate changes over Western Europe', *Climatic Change*, **10**, 1993, pp. 11–42.

14 From A. B. Pittock *et al.*, 1991, quoted in *Climate Change 1992: the Supplementary Report to the IPCC Assessment*, eds. J. T. Houghton, B. A. Callander and S. K. Varney, CUP, 1992, p. 120.

15 Studies of other stars are providing further information, see E. Nesme–Ribes *et al.*, *Scien. Amer.*; August 1996, pp. 31–6.

7 The Impacts of Climate Change

The last two chapters have detailed the climate change which we can expect next century because of human activities in terms of temperature and rainfall. To be useful to human communities, these details need to be turned into descriptions of the impact of climate change on human resources and activities. The questions to which we want answers are: how much will sea level rise and what effect will that have?; how much will water resources be affected?; what will be the impact on agriculture and food supply?; will natural ecosystems suffer damage and how will human health be affected? This chapter considers these questions.[1]

A complex network of changes

In outlining the character of the likely climate change in different regions of the world, the last chapter showed that it is likely to vary a great deal from place to place. For instance, in some regions precipitation will increase, in other regions it will decrease. Not only is there a large amount of variability in the character of the likely change, there is also variability in the sensitivity (for definition see box) of different systems to climate change. Different ecosystems, for instance, will respond very differently to changes in temperature or precipitation.

There will be a few impacts of the likely climate change which will be positive so far as humans are concerned. For instance, in parts of Siberia or northern Canada increased temperature will tend to lengthen the growing season with the possibility in these regions of growing a greater variety of crops.

Sensitivity, adaptability and vulnerability: some definitions

Sensitivity to climate change refers to the degree to which a system responds to climate change.

Adaptability to climate change refers to the degree to which adjustments are possible in practices, processes or structures of systems to projected or actual changes of climate.

Vulnerability to climate change refers to the extent to which climate change may damage a system. It depends both on a system's sensitivity and on its ability to adapt to a given change.

Both the magnitude and the rate of climate change are important in determining the sensitivity, adaptability and vulnerability of a system.

However, because, over centuries, human communities have adapted their lives and activities to the present climate, most changes in climate will tend to produce an adverse impact. If the changes occur rapidly, rapid and possibly costly adaptation to a new climate will be required by the affected community. An alternative might be for the affected community to migrate to a region where less adaptation would be needed – a solution which has become increasingly difficult or, in some cases, impossible in the modern crowded world.

As we consider the questions posed at the start of this chapter, it will become clear that the answers are far from simple. It is relatively easy to consider the effects of a particular change (in say, sea level or water resources) supposing nothing else changes. But other factors will change. Some adaptation, for both ecosystems and human communities, may be relatively easy to achieve; in other cases, adaptation may be difficult, very costly or even impossible. In assessing the effects of global warming and how serious they are, allowance must be made for response and adaptation. The likely costs of adaptation also need to be put alongside the costs of the losses or impacts connected with global warming.

The assessment of the impacts of global warming is also made more complex because global warming is not the only human induced environmental problem. The loss of soil and its impoverishment (through poor agricultural practice), the over-extraction of groundwater and the damage due to acid rain are examples of environmental degradations on local or regional scales which are having a substantial impact now[2]. If they are not corrected they will tend to exacerbate the negative impacts likely to arise from global warming.

For these reasons, the various effects of climate change so far as they concern human communities and their activities will be put in the context of other factors which might alleviate or exacerbate their impact. The following paragraphs will look at various impacts in turn and then bring them together in a consideration of the overall impact.

How much will sea level rise?

There is plenty of evidence for large changes in sea level during the Earth's past history. For instance, during the warm period before the onset of the last ice age, about 120,000 years ago, the global average temperature was a little warmer than today (Fig. 4.4). Average sea level then was about 5 or 6 metres higher than it is today. When ice cover was at its maximum towards the end of the ice age, some 18,000 years ago, sea level was over 100 m lower than today, sufficient, for instance, for Britain to be joined to the continent of Europe.

It is often thought that the main cause of these sea-level changes was the melting or growth of the large ice-sheets that cover the polar regions. It is certainly true that the main reason for the drop in sea level 18,000 years ago was the amount of water locked up in the large extension of the polar ice-sheets. In the northern hemisphere these extended in Europe as far south as southern England and in North America to south of the Great Lakes. It must also be true that the main reason for the 5 or 6 m higher sea level during the last warm interglacial period was a reduction in the Antarctic or Greenland ice-sheets. But changes over shorter periods are largely governed by other factors which combine to produce a significant effect on the average sea level.

Various contributions to the likely sea-level rise next century are shown in Fig. 7.1. The largest of these comes from thermal expansion of water in the oceans; as the oceans warm the water expands and the sea level rises (see box). The other main contribution comes from the melting of glaciers. If all glaciers outside Antarctica and Greenland were to melt, the rise in sea level would be about 50 cm (between 40 and 60 cm). Substantial glacier retreat has occurred during the past century and it is estimated that glacier melting has contributed between about 2 and 5 cm to the observed global sea-level rise of between 10 and 25 cm during this period. Modelling the effect of climate change on the behaviour of glaciers is, however, complex. The growth or decay of a glacier depends on the balance between the amount of snowfall on it, especially in winter, and the amount of melting in the summer. Both winter snowfall and average summer temperature are important, and both must be taken into account in future projections of the rate of glacier melting.

The uncertainties in the estimates in Fig. 7.1 are substantial. First of all there is the uncertainty in the actual temperature rise to be expected depending on the value chosen for the climate sensitivity (Fig. 6.2). Different models also give substantially different estimates of the amount of sea-level rise due to the melt from glaciers and small ice caps ranging from about 9 to 25 cm around the average of 16 cm. The total range of uncertainty by 2100 is from about 15 cm to about 1 metre.

It is interesting and perhaps surprising that the net contribution expected from changes in the Antarctic and Greenland ice-sheets is small. For both ice-sheets there are two competing effects[3]. In a warmer world, there is more water vapour in the atmosphere which leads to more snowfall. But there is also more ablation (erosion by melting) of the ice around the boundaries of the ice-sheets where melting of the ice and calving of icebergs occur during the summer months. For Antarctica, the estimates are that accumulation is greater than ablation, leading to a small net growth. For Greenland, ablation is greater than accumulation. For the two taken together the net effect is about zero, although there is considerable uncertainty in that estimate.

A part of the Antarctic ice-sheet in the west of Antarctica is often mentioned as of special concern. A large portion of it is grounded well below sea level.

Fig. 7.1
Estimates of sea-level rise next century for the IPCC IS 92a scenario of greenhouse gas emissions (I have called it the business-as-usual scenario), showing the contributions from various factors[4].

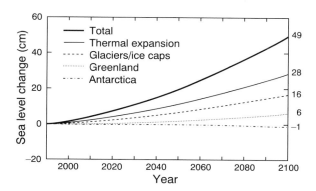

Thermal expansion of the oceans

A large component of sea-level rise is due to thermal expansion of the oceans. Calculation of the precise amount of expansion is complex because it depends critically on the water temperature. For cold water the expansion for a given change of temperature is small. The maximum density of sea water occurs at temperatures close to 0 ºC; for a small temperature rise at a temperature close to 0 ºC, therefore, the expansion is negligible. At 5 ºC (a typical temperature at high latitudes), a rise of 1 ºC causes an increase of water volume of about 1 part in 10,000 and at 25 ºC (typical of tropical latitudes) the same temperature rise of 1 ºC increases the volume by about 3 parts in 10,000. For instance, if the top hundred metres of ocean (which is approximately the depth of what is called the mixed layer) were at 25 ºC, a rise to 26 ºC would increase its depth by about 3 centimetres.

A further complication is that not all the ocean changes temperature at the same rate. The mixed layer fairly rapidly comes into equilibrium with changes induced by changes in the atmosphere. The rest of the ocean changes comparatively slowly (the whole of the top kilometre will, for instance, take decades to warm up); some parts may not change at all. To calculate the effect due to thermal expansion, therefore, it is necessary to employ the results of an ocean model – of the kind described in Chapter 5.

There have been suggestions that it could disintegrate rapidly, in which case its melting would cause sea level to rise by about 5 m. It is not known whether melting of the West Antarctic ice-sheet contributed to the rise in sea level during the last interglacial period, 120,000 years ago. Although scientists are not yet very confident in their ability to model the dynamic behaviour of large ice-sheets, there is no reason to suppose there is danger in the short term (for instance, during the next century) of the collapse of any of the major ice-sheets. Much greater understanding of the behaviour of large ice-sheets must be obtained before the amount of warming which might induce such collapse can be estimated.

According to the estimates for the business-as-usual scenario shown in Fig. 7.1, the total average sea-level rise is predicted to be about 12 cm by 2030 and about 50 cm by the year 2100. Because of the slow rate of rise in temperature of much of the oceans, sea-level rise resulting from global warming will lag behind temperature change at the surface. To illustrate this, Fig. 7.2 shows what would happen if the temperature rose as in the business-as-usual scenario until the year 2030, at which time greenhouse gas concentrations were stabilized so that no further change in radiative forcing of the climate would occur after that date. By 2030, sea level would have risen about 12 cm. During the remaining 70 years of the next century a further 18 cm or so of sea-level rise would be expected, the rise continuing at about the same rate for the following centuries as more of the ocean warms up.

These estimates of average sea-level rise provide a general guide as to what can be expected. Sea-level rise, however, will not be uniform over the globe.

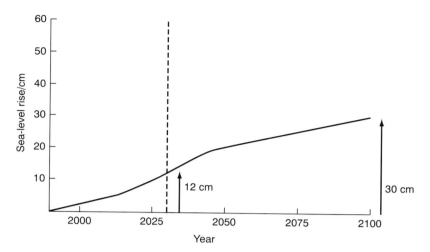

Fig. 7.2
Estimate of sea-level rise under the IPCC business-as-usual scenario of greenhouse gas emissions until the year 2030. An additional rise in sea level will occur during the remainder of the century even if climate forcing is stabilized in 2030[5].

The effects of thermal expansion in the oceans will vary considerably with location. Further, movements of the land occurring for natural reasons (for instance, the relief of stress in the Earth's mantle due to the removal of the last ice-sheets, which still continues), or because of human activities (for instance, the removal of groundwater), can have comparable effects to the rate of sea-level rise arising from global warming. At any given place, all these factors have to be taken into account in determining the likely value of future sea-level rise.

The impacts of sea-level rise

A rise in average sea level of 12 cm by 2030 and about half a metre by the end of the next century may not seem a great deal. Many people live sufficiently above the level of high water not to be directly affected. However, half of humanity inhabits the coastal zones around the world. Within these, the lowest lying are some of the most fertile and densely populated. To people living in these areas, even a fraction of a metre increase in sea level can add enormously to their problems. Some of the areas which are especially vulnerable are Bangladesh and similar delta areas, the Netherlands and the small low-lying islands in the Pacific and other oceans.

Bangladesh is a densely populated country of about 120 million people located in the complex delta region of the Ganges, Brahmaputra and Meghna Rivers. About 7 per cent of the country's habitable land (with about 6 million population) is less than 1 metre above sea level and about 25 per cent (with about 30 million population) is below the 3 metre contour (Fig. 7.3). Estimates of the sea-level rise next century are of about 1 metre by 2050 (compounded of 70 cm due to subsidence because of land movements and removal of ground-water and 30 cm from the effects of global warming) and nearly 2 m by 2100 (1.2 m due to subsidence and 70 cm from global warming)[6] – although there is a large uncertainty in these estimates.

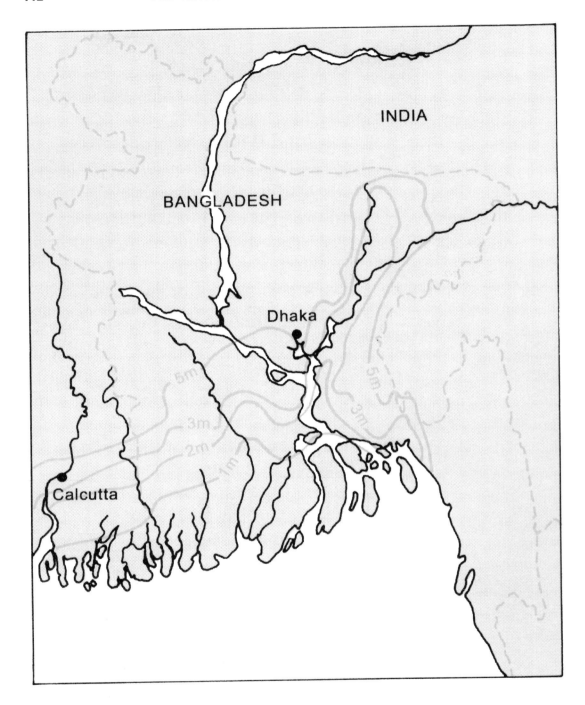

Fig. 7.3
Land affected in Bangladesh by various amounts of sea-level rise[7].

It is quite impractical to consider full protection of the long and complicated coastline of Bangladesh from sea-level rise. Its most obvious effect, therefore, is that substantial amounts of good agricultural land will be lost. This is serious: half the country's economy comes from agriculture and 85 per cent of the nation's population depends on agriculture for its livelihood. Many of these people are at the very edge of subsistence.

But the loss of land is not the only effect of sea-level rise. Bangladesh is extremely prone to damage from storm surges. Every year, on average, at least one major cyclone attacks Bangladesh. During the past twenty-five years there have been two very large disasters with extensive flooding and loss of life. The storm surge in November 1970 is probably the largest of the world's natural disasters in recent times; it is estimated to have claimed the lives of over a quarter of a million people. Well over a hundred thousand are thought to have lost their lives in a similar storm in April 1991. Even small rises in sea level add to the vulnerability of the region to such storms.

There is a further effect of sea-level rise on the productivity of agricultural land: the intrusion of saltwater into fresh groundwater resources. At the present time, it is estimated that in some parts of Bangladesh saltwater extends seasonally inland over 150 km. With a one metre rise in sea level, the area affected by saline intrusion could double, with a large effect on the supply of fresh water[8].

What possible responses can Bangladesh make to these likely future problems? Over the timescale of change which is currently envisaged it can be supposed that the fishing industry can relocate and respond with flexibility to changing fishing areas and changing conditions. It is less easy to see what the population of the affected agricultural areas can do to relocate or to adapt. No significant areas of agricultural land are available elsewhere in Bangladesh to replace that lost to the sea, nor is there anywhere else in Bangladesh where the population of the delta region can easily be located. It is clear that very careful study and management of all aspects of the problem is required. The sediment brought down by the rivers into the delta region is of particular importance. The amount of sediment and how it is used can have a large effect on the level of the land affected by sea-level rise. Careful management is therefore required upstream as well as in the delta itself; groundwater as well as sea defences must also be managed carefully if some alleviation of the effects of sea-level rise is to be achieved.

A similar situation exists in the Nile delta region of Egypt. The likely rise in sea level next century is made up from local subsidence and from global warming in much the same way as for Bangladesh – approximately 1 metre by 2050 and 2 metres by 2100. About 12 per cent of the country's arable land with a population of over 7 million people would be affected by a 1 metre rise of sea level[9]. Some protection from the sea is afforded by the extensive sand dunes but only up to half a metre or so of sea-level rise[10].

Many other examples of vulnerable delta regions, especially in south-east Asia and Africa, can be given where the problems would be similar to those in Bangladesh and in Egypt. For instance, several large and low-lying alluvial plains are distributed along the eastern coastline of China. A sea-level rise of just half a metre would inundate an area of about 40,000 square kilometres (about the area of the Netherlands)[11] where over 30 million people currently live. A particular delta which has been extensively studied is that of the Mississippi in North America. These studies underline the point that human activities and industry are already exacerbating the potential problems of sea-level rise due to global warming. Because of river management little sediment is delivered by the river to the delta to counter the subsidence occurring because of long-term movements of the Earth's crust. Also, the building of canals and dykes has inhibited the input of sediments from the ocean[12]. Studies of this kind emphasize the importance of careful management of all activities influencing such regions, and the necessity of making maximum use of natural processes in ensuring their continued viability.

We now turn to the Netherlands, a country more than half consisting of coastal lowlands, mainly below present sea level. It is one of the most densely populated areas in the world; 8 of the 14 million inhabitants of the region live in large cities like Rotterdam, The Hague and Amsterdam. An elaborate system of about 400 km of dykes and coastal dunes, built up over many years, protects it from the sea. Recent methods of protection, rather than creating solid bulwarks, make use of the effects of various forces (tides, currents, waves, wind and gravity) on the sands and sediments so as to create a stable barrier against the sea – similar policies are advocated for the protection of the Norfolk coast in eastern England[13]. Protection against sea-level rise next century will require no new technology. Dykes and sand dunes will need to be raised; additional pumping will also be necessary to combat the incursion of saltwater into freshwater aquifers. It is estimated[14] that an expenditure of about 10 thousand million dollars (US) would be required for protection against a sea-level rise of 1 metre.

The third type of area of especial vulnerability is the low-lying small island. Half a million people live in archipelagos of small islands and coral atolls, such as the Maldives in the Indian Ocean, consisting of 1,190 individual islands, and the Marshall Islands in the Pacific, which lie almost entirely within three metres of sea level. Half a metre or more of sea-level rise would reduce their areas and remove up to 50 per cent of their groundwater. The cost of protection from the sea is far beyond the resources of these islands' populations. For coral atolls, rise in sea level at a rate of up to about half a metre per century can be matched by coral growth, providing that growth is not disturbed by human interference and providing also that the growth is not inhibited by a rise in the maximum sea temperature exceeding about 1–2 $^{\circ}$C[15].

These are some examples of areas particularly vulnerable to sea-level rise. Many other areas in the world will be affected in similar, although perhaps less dramatic, ways. Many of the world's cities are close to sea level and are being increasingly affected by subsidence because of the withdrawal of groundwater. The rise of sea level due to global warming will add to this problem. There is no technical difficulty for most cities in taking care of these problems, but the cost of doing so must be included when calculating the overall impact of global warming.

So far, in considering the impact of sea-level rise, places of dense population where there is a large effect on people have been considered. There are also areas of importance where few people live. The world's wetlands and mangrove swamps currently occupy an area of about a million square kilometres (the figure is not known very precisely), equal approximately to twice the area of France. They contain much biodiversity and their biological productivity equals or exceeds that of any other natural or agricultural system. Over two-thirds of the fish caught for human consumption, as well as many birds and animals, depend on coastal marshes and swamps for part of their life cycles, so they are vital to the total world ecology. Such areas can adjust to slow levels of sea-level rise, but there is no evidence that they could keep pace with a rate of rise of greater than about 2 mm per year – 20 cm per century. What will tend to occur, therefore, is that the area of wetlands will extend inland, sometimes with a loss of good agricultural land. However, because in many places such extension will be inhibited by the presence of flood embankments and other human constructions, erosion of the seaward boundaries of the wetlands will lead more usually to a loss of wetland area. Because of a variety of human activities (such as shoreline protection, blocking of sediment sources, land reclamation, aquaculture development and oil, gas and water extraction) coastal wetlands are currently being lost at a rate of 0.5–1.5% per year. Sea level rise because of climate change would further exacerbate this loss[16].

To summarize the impact of the half metre or so of sea-level rise due to global warming which could occur next century: global warming is not the only reason for sea-level rise but it is likely to exacerbate the impacts of other environmental problems. Careful management of human activities in the affected areas can do a lot to alleviate the likely effects, but substantial adverse impacts will remain. In delta regions, which are particularly vulnerable, sea-level rise will lead to substantial loss of agricultural land and salt intrusion into freshwater resources. In Bangladesh, for instance, over 10 million people are likely to be affected by such loss. A further problem in Bangladesh and other low-lying tropical areas will be the increased intensity and frequency of disasters because of storm surges. Each year, the number of people worldwide experiencing flooding because of storm surges, estimated at about 45 million, would double with a 50 cm sea level rise. Low-lying small islands will also suffer loss of land and freshwater supplies. Countries like the Netherlands and many cities in coastal regions will have to spend substantial sums on protection against the sea. Significant amounts of land will also be lost near the important wetland areas of the world. Attempts to put costs against these impacts, in both money and human terms, will be considered later in the chapter.

In this section we have considered the impacts of sea-level rise for the next century. Because, as we have seen, the ocean takes centuries to adjust to an increase in surface temperature, the longer-term impacts of sea-level rise also need to be emphasized. Even if the concentrations of greenhouse gases in the atmosphere were stabilized so that anthropogenic climate change is halted, the sea level will continue to rise for many centuries as the whole ocean adjusts to the new climate.

The impact on fresh water resources

The global water cycle is a fundamental component of the climate system. Water is cycled between the oceans, the atmosphere and the land surface

(Fig. 7.4). Through evaporation and condensation it provides the main means whereby energy is transferred to the atmosphere and within it. Water is essential to all forms of life; the main reason for the wide range of life forms, both plant and animal, on the Earth is the extremely wide range of variation in the availability of water. In wet tropical forests, the jungle teems with life of enormous variety. In drier regions sparse vegetation exists, of a kind which can survive for long periods with the minimum of water; animals there are also well adapted to dry conditions.

Water is also a key substance for humankind; we need to drink it, we need it for the production of food, for health and hygiene, for industry and transport. Humans have learnt that the ways of providing for livelihood can be adapted to a wide variety of circumstances regarding water supply except, perhaps, for the completely dry desert. Water availability for domestic, industrial and agricultural use averaged per capita in different countries varies from 1,000 cubic metres (220,000 imperial gallons) per year to over 50,000 cubic metres (11 million imperial gallons)[18] – although quoting average numbers of that kind hides the enormous disparity between those in very poor areas who may walk many hours each day to fetch a few gallons and many in the developed world who have access to virtually unlimited supplies at the turn of a tap.

The demands of increased populations and the desire for higher standards of living have brought with them much greater requirements for fresh water.

Fig. 7.4
The global water cycle (in thousands of cubic kilometres per year)[17], showing the key processes of evaporation, precipitation, transport as vapour by atmospheric movements and transport from the land to the oceans by run-off or groundwater flow.

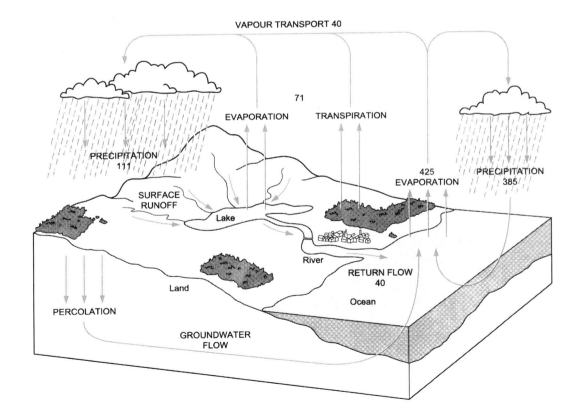

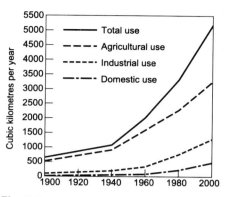

Fig. 7.5
Global water use for different purposes,
1900–2000, in cubic kilometres per year[19].

During the last 50 years water use worldwide has grown fourfold (Fig. 7.5); it now amounts to about 10 per cent of the estimated global total of the river and groundwater flow from land to sea (Fig. 7.4). Two-thirds of human water use is currently for agriculture, much of it for irrigation; about a quarter is used by industry; only 10 per cent or so is used domestically. Increasingly, water stored over hundreds or thousands of years in underground aquifers is being tapped for current use. With this rapid growth of demand comes greatly increased vulnerability regarding water supplies.

A further vulnerability arises because many of the world's major sources of water are shared. About half the land area of the world is within water basins which fall between two or more countries. There are 44 countries for which at least 80 per cent of their land areas falls within such international basins. The Danube, for instance, passes through 12 countries which use its water, the Nile water through 9, the Ganges-Bramaputra through 5. Other countries where water is scarce are critically dependent on sharing the resources of rivers such as the Euphrates and the Jordan. The achievements of agreements to share water often bring with them demands for more effective use of the water and better management. Failure to agree brings increased possibility of tension and conflict. The former United Nations Secretary-General, Boutros Boutros-Ghali has said that 'The next war in the Middle East will be fought over water, not politics'[20].

The availability of fresh water will be substantially changed in a world affected by global warming. The increase of temperature will mean that a higher proportion of the water falling on the Earth's surface will evaporate. This would not matter if there were more rainfall to make up for the evaporation. However, the projections presented in Chapter 6, for a business-as-usual scenario of carbon dioxide emissions next century, showed that some parts of the world would have less rainfall, especially in summer. The combined effect of less rainfall and more evaporation in these places would mean less run-off. Other parts of the world would have increased precipitation, possibly for instance the monsoon regions of south-east Asia, where studies indicate the possibility of substantially greater run-off in the summer (see box on p. 123).

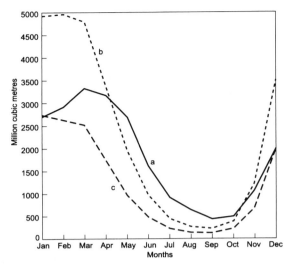

Fig. 7.6
Simulations of average monthly run-off in the Sacramento
basin of California comparing (a) current climate with (b)
changed climates with a 4 °C temperature increase and 20%
increase in rainfall and (c) with the same temperature
increase but with 20% decrease of rainfall[21].

The run-off in rivers and streams is what is left from the precipitation which
falls on the land after some has been taken by evaporation and by transpiration
from plants; it is the major part of what is available for human use. The amount
of run-off is highly sensitive to changes in climate; even small changes in
the amount of precipitation or in the temperature (affecting the amount of
evaporation) can have a big influence on it. To illustrate this Fig. 7.6 shows
simulations, carried out for the Sacramento Basin in California, USA (a region
where water is stored for some of the year in mountain snow), of changes in
run-off with changed climate conditions. With a 4 °C regional temperature rise
and 20 per cent decreased rainfall, the run-off in the summer months falls to
between 20 per cent and 50 per cent of its normal value. Even with 20 per cent
increased rainfall and the same temperature increase, summer run-off still
remains well below normal. Watersheds in arid or semi-arid regions are especially sensitive because the annual run-off is in any case highly variable.

Some watersheds in mid-latitudes in the northern hemisphere, where
snowmelt is an important source of run-off, can also be severely affected. For
these places, as temperatures rise, winter run-off will increase substantially and
spring high water will be much reduced. Further, as we saw earlier in the chapter up to one half of the mass of mountain glaciers and small ice caps may melt
away over the next hundred years which could substantially change the seasonal distribution of river flow and water supply for hydroelectric generation
and agriculture.

So far when mentioning changes in temperature or rainfall it is changes in
the average with which we have been concerned; for instance, the simulations
in Fig. 7.6 are for average conditions. But, as has been constantly emphasized,

the severity of climate impacts depends to a great degree on extreme conditions. This is well illustrated by looking at the scale of natural disasters involving water – either too much water in floods or too little in droughts. In the list of natural disasters listed in Table 1.1, floods were third (after tropical cyclones and earthquakes) in order of severity leading to loss of life. A Unesco study in 1973 estimated that each year, in Asia alone, river floods damage or destroy about 40,000 square kilometres of land and crops, and affect the lives and well-being of over 17 million people. During the period 1980–85, more than 160 major floods were recorded, killing or injuring more than 120,000 people, destroying the homes of nearly 20 million and causing damage estimated at over 20 thousand million US dollars – a very large amount indeed for the economies of the countries involved. Droughts do not appear high up on the table of natural disasters, not because they are unimportant, but because, unlike most other disasters, their effects tend to be felt over a long period of time. The 'dust bowl' years in the 1930s in the United States are still within living memory, as are the droughts and famines in India in 1965–67 which, it is estimated, claimed one and a half million lives. We have all been made very aware of the disastrous consequences of drought in the Sahel region and in other parts of Africa in the 1980s – which are still recurring only too frequently on that continent.

Any temperature or rainfall record shows a large variability. The inevitable result of variability added to higher average temperatures (meaning higher evaporation) and higher average rainfall will be a greater number and greater intensity of both droughts and floods[22]. Some of the areas likely to be affected are just those areas which are particularly vulnerable to floods and droughts at the moment. But, as was also implied in Chapter 6, droughts and floods are also likely to occur in locations where, at present, such disasters are rare.

We can identify watersheds that are particularly vulnerable to climate change by asking certain questions about them[23].

- How much water storage is there in the watershed relative to the annual flow? In Colorado, in the United States, for instance, the storage is four times the annual flow, whereas in the Atlantic States it is only one tenth of the annual flow.
- How large is the demand as a percentage of the potential supply? Again in the United States, for the Missouri river basin it is 30 per cent, for the Rio Grande it is 64 per cent and for the lower Colorado 96 per cent. Almost none of the water in the Colorado river reaches the sea. Therefore, though the Colorado has substantial storage and is therefore not very sensitive to annual variations, the amount of use in its lower reaches means that, over a number of years, any reduction of its flow is bound to imply lower water availability.
- How much groundwater is being used? There are many places in the world where groundwater is being used faster than it is being replenished. To give two examples, for more than half the land area of the United States over a quarter of the groundwater withdrawn is not being replenished, so every year the water has to be extracted from deeper levels; and in Beijing in China the water table is falling by two metres a year as its groundwater is pumped out.

Study of the 'MINK' region in the United States

The United States Department of Energy has carried out a detailed study[24] of the likely effects of climate change on a region (known as the MINK region) in the centre of the United States comprising the states of Missouri, Iowa, Nebraska and Kansas. Included within the region are parts of four major river basins – the Missouri, the Arkansas, the Upper and the Lower Mississippi. Water is already a scarce resource within the MINK states; much of the area's irrigation relies on non-renewable groundwater supplies. These will diminish with time, so that even in the absence of climate change less water will be available, especially for irrigation.

To provide an analogue of the climate which might be expected with increased carbon dioxide, the period of the 1930s was chosen, when the average temperature in the region was about 1 °C warmer than in the period 1950–1980 (the 'control' period) and the average precipitation about 10 per cent lower than in the control period.

Water would become scarcer under the analogue climate compared with the control[25]. The hotter and drier conditions would increase evaporation and reduce run-off. Streamflow would drop by about 30 per cent in the Missouri and the Upper Mississippi basins and by about 10 per cent in the Arkansas. Most streams would fall well short of supplying both the desired instream flows and the current levels of consumption use.

Under the analogue climate, irrigated agriculture would be bound to decline substantially because of the increased constraints on groundwater use coupled with less water availability from other sources. This would also result in a drive to increased efficiency, albeit at greater cost. Maintaining the high priority currently given to navigation on the main stem of the Missouri would become very costly.

• How variable are the stream and river flows? This question is particularly relevant to arid and semi-arid areas. Detailed studies taking these criteria into account have been carried out for a number of areas; one example for the MINK (Missouri, Iowa, Nebraska and Kansas) region of the United States is shown in the box.

There is another reason, not unconnected with global warming, for the vulnerability of water supplies: the link between rainfall and changes in land use. Extensive deforestation can lead to large changes in rainfall (see box). A similar tendency to reduced rainfall can be expected if there is a reduction in vegetation over large areas of semi-arid regions. Such changes can have a devastating and widespread effect and assist in the process of desertification. This is a potential threat to the drylands covering about one quarter of the land area of the world (see box on p. 122).

What sort of action can be taken to decrease the vulnerability of human communities to water supplies? Irrigation accounts for about two-thirds of world water use, and is of great importance to world agriculture. Irrigation is applied to about one-sixth of the world's farmland which produces about one-third of the world's crops. In some areas the ratio is much higher; for instance, over 80 per cent of the agricultural land in California is irrigated. Most irrigation is through open ditches, which is very wasteful of water; over 60 per cent is lost through evaporation and seepage. Microirrigation techniques, in which perforated pipes deliver water directly to the plants, provide large opportunities for water conservation, making it possible to expand irrigated fields without

Deforestation and changes in rainfall

Changes in land use such as those brought about by deforestation can have an effect on the amount of rainfall, for three main reasons. Over a forest there is a lot more evaporation of water (through the leaves of the trees) than there is over grassland or bare soil, hence the air will contain more water vapour. Also, a forest reflects 12–15 per cent of the sunlight which falls on it, whereas grassland will reflect about 20 per cent and desert sand up to 40 per cent. A third reason arises from the roughness of the surface where vegetation is present.

An American meteorologist, Professor Jules Charney suggested in 1975, in the context of the drought in the Sahel, that there could be an important link between changes of vegetation (and hence changes of reflectivity) and rainfall. The increased energy absorbed at the surface when vegetation is present and the increased surface roughness both tend to stimulate convection and other dynamic activity in the atmosphere which leads to the production of rainfall.

Experiments with numerical models which include these physical processes demonstrate the effect. They indicate a reduction of about 15 per cent in rainfall if the forest north of 30ºS in South America were removed and replaced by grassland[26]. Similar model experiments for Zaire over a smaller region show an average reduction in rainfall of over 30 per cent[27]. A much more drastic experiment in which the Amazonian forest was removed and replaced with a desert surface showed a reduction in rainfall by 70 per cent to levels similar to that of the semi-arid regions of the Sahel part of Africa[28]. Such a model experiment does not represent a realistic situation, but it illustrates the significant impact that widespread deforestation could have on the local climate.

building new dams[29]. Management of the existing infrastructure can be improved, for instance by arranging for the integration of different supplies, and conservation in the domestic and industrial sectors can be encouraged. Most of these actions will cost money, although they may be much more cost-effective ways of coping with future change in water resources than attempting to develop major new facilities[30].

In summary, what are the likely effects of global warming on water supplies? Firstly, the current vulnerability of many communities to water shortage should be noted. This is especially true of arid and semi-arid regions where the increasing demands of human communities mean that droughts, even for short periods, are more disastrous than before. Vulnerability is well demonstrated in many areas of the world where the amounts of groundwater extraction greatly exceed its replenishment – a situation which cannot continue for very long into the future. Because of population growth these vulnerabilities will increase and will exacerbate the negative effects of global warming.

Secondly, climate change because of global warming will result in large changes in water supplies in many places. Although the present state of knowledge regarding regional and local climate change does not allow scientists to identify clearly the most vulnerable areas, they are able to indicate the sort of

Desertification

Drylands (defined as those areas where precipitation is low and where rainfall typically consists of small, erratic, short, high-intensity storms) cover about 40 per cent of the total land area of the world and support over one fifth of the world's population. Desertification in these drylands is the degradation of land because of decreased vegetation, reduction of available water, reduction of crop yields and erosion of soil. It results from excessive land use generally because of increased population, increased human needs, or political or economic pressures (for instance, the need to grow cash crops to raise foreign currency). It is often triggered or intensified by a naturally occurring drought.

The rate of desertification is currently about 60,000 square kilometres per year or 0.1 per cent per year of the total area of drylands[31]. It is a potential threat to 70 per cent of these drylands – over 25 per cent of the world's land area. Figure 7.7 shows how these arid areas are distributed over the continents.

Fig. 7.7
The world's drylands, by continent. The total area of drylands is about 60 million square kilometres (about 40% of the total land area), of which 10 million are hyper-arid deserts[32].

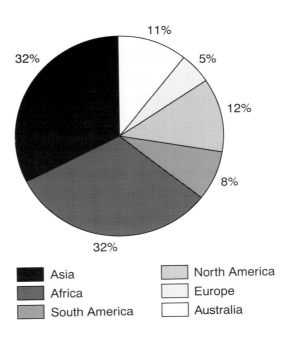

Legend	
■ Asia	▨ North America
▨ Africa	□ Europe
▨ South America	□ Australia

area which will be most affected. Such areas are those arid and semi-arid areas with reduced rainfall leading to greater aridity and even desertification; continental areas where decreased summer rainfall and increased temperature result in a substantial loss in soil moisture and much increased vulnerability to drought; and areas where the possibility of increased rainfall could lead to a greater incidence of floods. The changing pattern of climate extremes, especially droughts and floods, will be the cause of most of the problems. It is also the case that regions such as south-east Asia that are dependent on unregulated river systems are more sensitive to change than regions such as western Russia and the western United States that have large, regulated water resource

The monsoon regions of south-east Asia

A study for south-east Asia of the change in water resources which might occur with global warming has been carried out by Murari Lal from the Centre for Atmospheric Sciences in New Delhi[33]. He has used projections of future climate from the coupled atmosphere-ocean model of the Max Planck Institute in Hamburg, which has been shown to produce good simulations of the precipitation patterns for the region under the current climate.

For the business-as-usual increase in greenhouse gas emissions and assuming no significant increase in aerosols, the model results indicate for the next hundred years (up to about 2080) an average surface temperature increase over the land areas of about 3 °C, with increased precipitation and run-off in the flood-prone areas of north-east India and south China during the summer (Fig. 7.8). Lal points out the implications that such changes would have for water management and for land management, especially the control of erosion[34].

Other runs of the model demonstrated the sensitivity of the changes to the presence of aerosols[35]. When a substantial increase in sulphate aerosols was assumed, precipitation in north-east India decreased rather than increased.

Fig 7.8
Changes in average seasonal precipitation and run-off for the land areas of south-east Asia simulated by the Hamburg climate model[36].

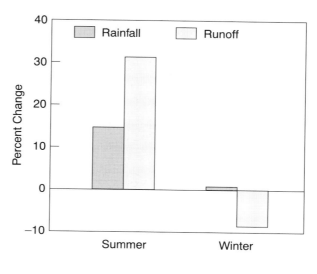

systems. Thirdly, some alleviating actions that can be taken have been described; the possible cost of these will be considered later in the chapter.

Impact on agriculture and food supply

Every farmer understands the need to grow crops or rear animals which are suited to the local climate. The distribution of temperature and rainfall during the year are key factors in making decisions regarding what crops to grow. These will change in the world influenced by global warming. The patterns of what crops are grown where will therefore also change. But these changes will be complex; economic and other factors will take their place alongside climate change in the decision-making process.

There is enormous capacity for adaptation in the growth of crops for food – as is illustrated by what was called the Green Revolution of the 1960s when the development of new strains of many species of crops resulted in large increases in productivity. Between the mid-1960s and the mid-1980s global food production rose by an average annual rate of 2.4 per cent – faster than global population – more than doubling over that thirty-year period. Grain production grew even faster, at an annual rate of 2.9 per cent[37].

With the detailed knowledge of the conditions required by different species and the expertise in genetic manipulation available today, there should be little difficulty in matching crops to new climatic conditions over large parts of the world. At least, that is the case for crops that mature over a year or two. Forests reach maturity over much longer periods, from decades up to a century or even more. The projected rate of climate change is such that, during this time, trees may find themselves in a climate to which they are far from suited. The temperature regime or the rainfall may be substantially changed, resulting in stunted growth or a greater susceptibility to disease and pests. The impact of climate change on forests is considered in more detail in the next section.

An example of adaptation to changing climate is the way in which farmers in Peru adjust the crops they grow depending on the climate forecast for the year[38]. Peru is a country whose climate is strongly influenced by the cycle of El Niño events described in Chapters 1 and 5. Two of the primary crops grown in Peru, rice and cotton, are very sensitive to the amount and the timing of rainfall. Rice requires large amounts of water; cotton has deeper roots and is capable of yielding greater production during years of low rainfall. In 1983, following the 1982–83 El Niño event, agricultural production dropped by 14 per cent. By 1987 forecasts of the onset of El Niño events had become sufficiently good for Peruvian farmers to take them into account in their planning. In 1987, following the 1986–87 El Niño, production actually increased by 3 per cent, thanks to a useful forecast.

The availability of water is the most important of the factors affecting agriculture and food production. The vulnerability of water supplies to climate change carries over into a vulnerabilty in the growing of crops and the production of food. Thus the arid or semi-arid areas, mostly in developing countries, are most at risk. A further factor, which can actually lead to increased production as a result of climate change, is the boost to growth which is given, particularly to some crops, by increased atmospheric carbon dioxide (see box).

Detailed studies have been carried out of the sensitivity to climate change of the major crops which make up a large proportion of the world's food supply (for an example see box). They have used the results of models to estimate the changes in temperature and precipitation which might be expected when the atmosphere has a doubled carbon dioxide concentration. They have also included the possible effects of economic factors and the possibility of modest levels of adaptation. Such models are at an early stage of development and cannot be employed as detailed predictions. They tend to show that, with appropriate adaptation, the effect of the average climate change which would occur with doubled carbon dioxide on the supply of food for the world as a whole is not likely to be very large. What has not been studied to any extent is the likely effect of climate extremes (especially of the incidence of drought)

The carbon dioxide 'fertilization' effect

An important positive effect of increased carbon dioxide concentrations in the atmosphere is the boost to growth in plants given by the additional carbon dioxide. Higher CO_2 concentrations stimulate photosynthesis, enabling the plants to fix carbon at a higher rate. This is why in glasshouses additional CO_2 may be introduced artificially to increase productivity. The effect is particularly applicable to what are called C3 plants (such as wheat, rice and soya bean), but less so to C4 plants (for example maize, sorghum, sugar-cane, millet and many pasture and forage grasses). Under ideal conditions it can be a large effect (for doubled CO_2, up to 25 per cent for wheat and rice and up to 40 per cent for soybean)[39]. However, under real conditions on the large scale where water and nutrient availability are also important factors influencing plant growth, the increases, although difficult to estimate or measure, are likely to be substantially less than the ideal[40].

---

on global food supply. Nor have the studies considered the integrity of the world's soils, which are currently being lost by erosion at an alarming rate[41]. Further, a serious issue exposed by the studies is that climate change is likely to affect countries very differently. Production in developed countries with relatively stable populations may well increase, whereas that in many developing countries (where large increases in population are occurring) is likely to decline as a result of climate change. The disparity between nations will tend to become larger, as will the number of those at risk from hunger.

In looking to future needs, two activities which can be pursued now are particularly important. Firstly, there is large need for technical advances in agriculture in developing countries requiring investment and widespread local training. In particular, there needs to be continued development of programmes for crop breeding and management, especially in conditions of heat and drought. These can be immediately useful in the improvement of productivity in marginal environments today. Secondly, as was seen earlier when considering fresh water supplies, improvements need to be made in the availability and management of water for irrigation, especially in arid or semi-arid areas of the world.

Now to summarize the likely effect of global warming on agriculture and food supplies: although studies of the sensitivity of crop production and food supply are at an early stage, there is yet no strong evidence that the effect of climate change on global food supply is likely to be large. What needs urgent research is how well world agriculture will respond to extremes, such as prolonged droughts. Careful studies of this should be carried out as soon as possible.

It is fairly clear, however, that an increasing disparity is likely to develop between developed and developing countries. The surplus of food in developed countries is likely to increase, while developing countries will face large population increases coupled with a likely relative decrease in food production. Such a situation will raise enormous problems and serious deprivation especially in the developing world.

Climate change and world food supply

A detailed study of the effect of climate change on world food supply has been carried out at the Environmental Change unit at Oxford University with assistance from agricultural scientists in 18 countries[42]. For the details of climate change the results from several different models (of the kind described in Chapter 5) were used for the climate under doubled CO_2 concentration. Potential changes in national grain crop yields were estimated for wheat, rice, maize (which between them account for about 85 per cent of world cereal exports) and soybean (which accounts for about two-thirds of the trade in protein cake equivalent). These national crop yield changes were extrapolated to provide yield change estimates for other countries and other crops.

Various assumptions were made about the degree of adaptation by farmers. A world food trade model was employed to allow for possible levels of population growth and economic change.

On the assumption that the climate changes being estimated occur by the year 2060, assuming also a continuation in current trends in economic growth rates, partial trade liberalization and medium population growth rates, the main findings of the study are:

- The negative effects of climate change are to some extent compensated for by increased productivity due to the fertilization effect of increased CO_2 (see previous box) assumed in their model. With a modest level of farm-level adaptation (for example minor shifts in planting dates, changes in crop variety) the net effect of climate change is to reduce global cereal production from what it would be without climate change, by up to 5 per cent. This reduction can be overcome by more major forms of adaptation such as the installation of irrigation.
- Climate change would increase the disparities in cereal production between developed and developing countries. In the model, production in the developed world tended to benefit from climate change (possible increase of around 5 per cent) whereas production in developing nations declined (by around 10 per cent) as a result of climate change. Adaptation at the farm level does little to reduce these disparities.
- Cereal prices and thus the population at risk of hunger in developing countries are likely to increase despite adaptation.

The authors emphasize that, although the models and the methods they have employed are comparatively complex, there are many factors which have not been taken into account. For instance, they have not considered the availability of water supplies for irrigation. Further, (see Chapter 6) scientists are not yet very confident in the regional detail of climate change. The results, therefore, although giving a general indication of the changes that could occur, should not be treated as a detailed prediction. They highlight the importance of studies of this kind as a guide to future action.

One of these problems will be that of employment. Agriculture is the main source of employment in developing countries; people need employment to be able to buy food. With changing climate, as some agricultural regions shift, people will tend to attempt to migrate to places where they might be employed in agriculture. With the pressures of rising populations such movement is likely to be increasingly difficult and we can expect large numbers of environmental refugees.

The impact on natural ecosystems

A little over 10 per cent of the world's land area is under cultivation. The rest is to a greater or lesser extent unmanaged by humans. Of this about 30 per cent is natural forest. The variety of plants and animals which constitute a local ecosystem is sensitive to the climate, the type of soil and the availability of water. Ecologists divide the world into biomes – regions characterized by their distinctive vegetation. This is well illustrated by information about the distribution of vegetation over the world during past climates (eg for the part of North America shown in Fig. 7.9), which indicates what species and what ecosystems are most likely to flourish under different climatic regimes.

Changes in climate alter the suitability of a region for different species, and change their competitiveness within an ecosystem, so that even relatively small changes in climate will lead, over time, to large changes in the composition of an ecosystem. Since climate is the dominant factor determining the distribution of biomes (Fig. 7.10), information gleaned from paleo sources could be used to produce maps of the optimum distribution of natural vegetation under the climate scenarios expected to occur with global warming.

However, changes of the kind illustrated in Fig. 7.9 took place over thousands of years. With global warming similar changes in climate occur over a few decades. Most ecosystems cannot respond or migrate that fast. Natural ecosystems will therefore become increasingly unmatched to their environment. How much this matters will vary enormously from species to species; some are much more vulnerable to changes in average climate or climate extremes than others. But all will become more prone to disease and attack by pests. Any positive effect from added 'fertilization' due to increased carbon dioxide is likely to be more than outweighed by negative effects from other factors.

Trees are long-lived and take a long time to reproduce. Because it is not easy for them to respond quickly to climate change, the world's forests are likely to be the most affected of the world's ecosystems. Since also forests cover about a quarter of the world's total land area, the impact of climate change on the world's forests is particularly important.

Many trees are surprisingly sensitive to the average climate in which they develop. The environmental conditions (eg temperature and precipitation) under which a species can exist and reproduce is known as its niche. Climate niches for some typical tree species are illustrated in Fig 7.11; under some conditions a change as small as 1 °C in annual average temperature can make a substantial difference to a tree's productivity. For the likely changes in climate next century, a substantial proportion of existing trees will be subject to unsuitable climate conditions. This will be particularly the case in the boreal forests of the northern hemisphere where, as trees become less healthy, they will be more prone to pests, die-back and forest fires. One estimate projects that, under a doubled CO_2 scenario, up to 65 per cent of the current boreal forested area could be affected[43].

Fig. 7.9
Vegetation maps of the
south-eastern United
States during past
climate regimes[44]:
(a) for 18,000 years ago
at the maximum extent
of the last ice age,
(b) for 10,000 years
ago, (c) for 5,000 years
ago when conditions
were similar to the
present. A vegetation
map for 200 years ago
is similar to that in (c).

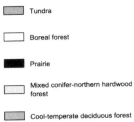

Tundra

Boreal forest

Prairie

Mixed conifer-northern hardwood
forest

Cool-temperate deciduous forest

Warm-temperate southeastern
evergreen forest

Sand dune scrub

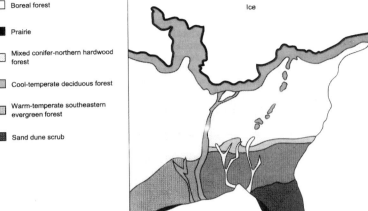

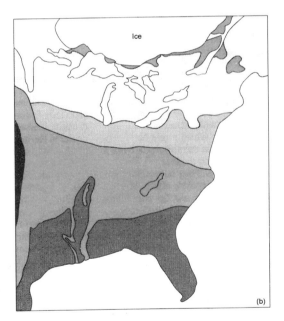

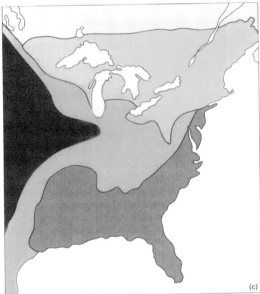

Fig. 7.10
The pattern of world
biome types related
to mean annual
temperature and
precipitation. Other
factors, especially the
seasonal variations of
these quantities, affect
the detailed distribution
patterns (after Gates[45]).

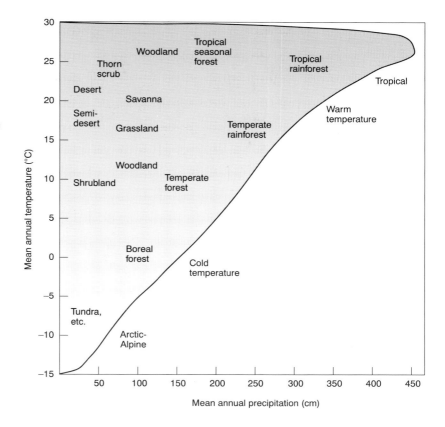

A decline in the health of many forests in recent years has received a great deal of attention, especially in Europe and North America where much of it is thought to be due to acid rain and other pollution originating from heavy industry, power stations and motor cars. Not all damage to trees, however, is thought to have this origin. Studies in several regions of Canada, for instance, indicate that the die-back of trees there is related to changes in climatic conditions, especially to successions of warmer winters and drier summers[46]. In some cases it may be the double effect of pollution and climate stress causing the problem; trees already weakened by the effects of pollution fail to cope with climate stress when it comes. The assessment of the impact of climate change carried out for the MINK region of the United States (see box on p. 120) concluded that, under the warmer, drier conditions of the analogue climate they studied, decline and die-back of the forested part of the region would reduce the mass of timber in the forest by 10 per cent over twenty years[47]. The results of these studies are indicative of the more serious levels of forest die-back which are likely to occur with the rapid rate of climate change expected with global warming. These stresses on the world's forests due to climate change will be concurrent with other problems associated with forests, in particular those of continuing tropical deforestation and of increasing demand for wood and wood products resulting from rapidly increasing populations especially in developing countries.

Arolla Pine

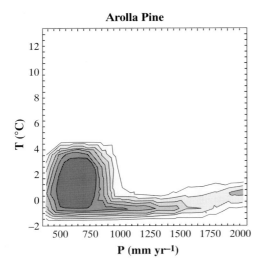

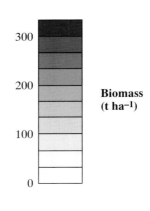

Biomass
(t ha^{-1})

Norway Spruce

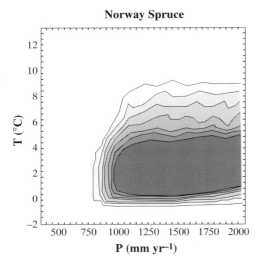

Common Beech

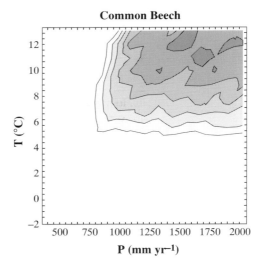

Fig 7.11
Simulated environmental realized niches (the realized niche describes the conditions under which the species is actually found) for three tree species, Arolla Pine, Norway Spruce and Common Beech[48]. Plots are of biomass generated per year against annual means of temperature and precipitation. Arolla Pine is a species with a particularly narrow niche. The narrower the niche, the greater the potential sensitivity to climate change.

If a stable climate is eventually reestablished, given adequate time (which could be centuries) different trees will be able to find again at some location their particular climatic niche. It is during the period of rapid change that most trees will find themselves unsatisfactorily located from the climate point of view.

It was mentioned in Chapter 3 that forests represent a large store of carbon; 80 per cent of above-ground and 40 per cent of below-ground terrestrial carbon is in forests. We also saw in Chapter 3 that tropical deforestation due to human activities is probably releasing between 1 and 2 Gt of carbon into the atmosphere each year. If, because of the rate of climate change, substantial stress and die-back occurs especially in boreal forests a release of carbon will occur. Just how large this will be is uncertain but estimates as high as 240 Gt over the next century have been quoted[49]. This potential positive feedback was mentioned in Chapter 3 (see the box on page 31); it could significantly increase the amount of carbon dioxide in the atmosphere.

A further concern about natural ecosystems, especially forests, relates to the diversity of species which they contain. Much concern is being expressed at the moment because of the threat to biodiversity from extensive deforestation. There is a danger that the rapid climate change expected as a result of global warming will also lead to significant loss of species in some ecosystems – although the likely effect is difficult to quantify.

So far we have been considering ecosystems on land. What about those in the oceans; how will they be affected by climate change? Although we know much less about ocean ecosystems, there is considerable evidence that biological activity in the oceans has varied during the cycle of ice ages. Chapter 3 noted (see box on p. 28) the likelihood that it was these variations in marine biological activity which provided the main control on atmospheric carbon dioxide concentrations during the past million years (see Fig. 4.4). The changes in ocean water temperature and the possible changes in some of the patterns of ocean circulation are likely to result in changes in the regions where upwelling occurs and where fish congregate. Some fisheries could collapse and others expand. At the moment the fishing industry is not well adapted to address major change[50].

The impact on human health

Human health is dependent on a good environment. Many of the factors that lead to a deteriorated environment also lead to poor health. Pollution of the atmosphere, polluted or inadequate water supplies, and poor soil (leading to poor crops and inadequate nutrition) all present dangers to human health and wellbeing and assist the spread of disease. As has been seen so far in considering the impacts of global warming, many of these factors will tend to be exacerbated through the climate change which will occur in the warmer world. The greater likelihood of extremes of climate, such as droughts and floods, will also bring increased risks to health from a variety of causes.

How about direct effects of the climate change itself on human health[51]? Humans can adapt themselves and their buildings so as to live satisfactorily in very varying conditions and have great ability to adapt to a wide range of climates. However, extreme conditions are not very tolerable and can encourage the transmission of some diseases. The main difficulty in assessing the impact of climate change on health is that of unravelling the influences of climate from the large number of other factors (including other environmental factors) that affect health.

The main direct effect of climate change on humans themselves will be that of heat stress in the extreme high temperatures that will become more frequent and more widespread. Studies using data from large cities where heat waves commonly occur show death rates which can be doubled or tripled during days of unusually high temperatures[52]. Although such an episode may be followed by a period when fewer deaths occur, showing that some of the deaths would in any case have occurred about that time, most of the increased mortality seems to be directly associated with the excessive temperatures. It might be thought that compensation for the periods of excessive heat would be provided by fewer occasions of severe cold. Studies tend to show, however, that increased mortality due to periods of excessive heat will considerably exceed any reduction in the periods of cold, especially amongst old people who find it particularly difficult to cope with high temperatures[53].

A further likely impact of climate change on health is the increased spreading of diseases in a warmer world. Many insect carriers of disease thrive better in warmer and wetter conditions. For instance, epidemics of diseases such as viral encephalitides carried by mosquitoes are known to be associated with the unusually wet conditions which occur in the Australian, American and African continents associated with different phases of the El Niño cycle[54]. Some diseases, currently largely confined to tropical regions, with warmer conditions could spread into mid-latitudes. Malaria is an example of such a disease which is spread by mosquitoes under conditions which are optimum in the temperature range of 15–32 °C with humidities of 50–60 per cent. It currently represents a huge global public health problem, causing annually around 350 million new infections and 2 million deaths. Estimates have been made indicating that the geographical zone of potential malaria transmission in response to world temperature increases at the upper part of the IPCC projected range (3–5 °C by 2100) could increase from approximately 45 per cent of the world population to approximately 60 per cent by the latter half of the next century – the increase in malaria incidence would be primarily in tropical, sub-tropical and less well-protected temperate-zone populations. Other diseases which are likely to spread for the same reason are yellow fever, dengue fever and some viral encephalitis.

The potential impact of climate change on human health could be large. However, the factors involved in all cases are highly complex; any quantitative conclusions will require a great deal more careful study of the direct effects of climate on humans and of the epidemiology of the diseases which are likely to be particularly affected. An international project called Global Health Watch has been proposed for the collection of the data required for such studies[55].

Costing the impacts

It is not an easy task to estimate the likely cost to the world community of the various impacts listed above. For a costing to be at all realistic, especially when it is to apply to periods of decades into the future, it must account not only for direct damage but also for the possibilities of adaptation. However, even crude attempts at costing help to provide an idea of the size of the problem.

The most detailed cost studies of the impacts of climate change in a world with carbon dioxide concentration doubled from its pre-industrial level have been carried out for the United States (see box). For those impacts against which some value of damage can be placed, estimates fall in the range 61–74 thousand million dollars per annum or between 1.1 and 1.5 per cent of the US Gross Domestic Product (GDP) in 1990. For other countries in the developed world, estimates of the cost of impacts in terms of percentage of GDP are similar; for the developing world, estimates of annual cost are typically around 5 per cent of GDP (with a range from 2 to 9 per cent of GDP[56]). Aggregated over the world the estimates are between 1.5 and 2 per cent of globally aggregated GDP (sometimes called Global World Product or GWP).

As the authors of these economic studies explain, their estimates are crude, are based on very broad assumptions and should not be considered as precise values. In particular they have mostly been calculated in terms of the impact on today's economies. Future impacts will depend on economic, demographic, environmental and other developments which will make future conditions very different from those today. However, they provide us with a first indication of the scale of the problem in economic terms. Later chapters will compare them with the cost of taking action to slow the onset of global warming or reduce its overall magnitude.

An initial inspection of the figures in Table 7.1 might suggest that the costs of global warming, although large, at a per cent or two of Gross World Product (GWP) are not impossibly large, and that we could probably buy our way out of the problems of its impact. There are, however, two factors which have been omitted from the studies cited above. The first is that the studies have only been concerned with the impacts of global warming up to about the middle of the next century when, for a business-as-usual scenario, the equivalent carbon dioxide concentration will have doubled from its pre-industrial value. The longer-term impacts, if the growth of greenhouse gases continues under business-as-usual, are likely to be much greater[57]. The second factor is that not all impacts can be quantified in terms of economic costs. For instance, the loss of human amenity, natural amenity or the loss of species cannot be easily expressed in money terms. This can be illustrated by focusing on those who are likely to be particularly disadvantaged by global warming. Most of them will be in the developing world at around the subsistence level. They will find their land is no longer able to sustain them because it has been lost either to sea-level rise or to extended drought. They will therefore wish to migrate and will become environmental refugees.

Estimating the costs of global warming

Most estimates of the costs of global warming have been carried out by estimating the annual cost of the damage which would be inflicted on current economies supposing the impacts were to be immediately realized of a change to an equilibrium climate with carbon dioxide concentration doubled from the pre-industrial period (i.e. a global average temperature rise of 2.5°C).

The most detailed studies so far have been carried out for the United States and are summarized in Table 7.1 against the various areas of impact presented in this chapter. For each impact area the range of estimates is given from three studies[58] which provide this sort of detail and which are based on the IPCC 'best' estimate of a 2.5°C global average temperature rise for doubled carbon dioxide. The range given for the total is that of the respective totals of the three studies.

Figures in the range 61–74 thousand million dollars represent between 1.1 and 1.5 per cent of the GDP of the United States at 1990 prices. Studies[59] of the costs arising from the likely impacts in the rest of the world come up with similar figures in terms of percentage of GDP in developed countries but with higher figures, typically around 5 per cent of GDP, for developing countries. For the region of the former Soviet Union, estimates of damage are somewhat less, typically less than 1 per cent of GDP.

All those who have carried out such studies emphasize their preliminary nature and the uncertainties involved in many of the estimates. On the one hand, many of the likely impacts (eg droughts, floods, loss of land, the migration of refugees, loss of species) are not well quantified; on the other, adaptation over the likely period of change is not fully taken into account. However, many of the estimates are considered to be deliberately conservative[60].

Table 7.1
Estimates in thousands of millions of US dollars (1990 value) per year of the costs of the impacts of climate change on the United States.

Loss of land and dyke construction due to sea-level rise	7–9
Loss of water supplies	7–16
Losses in the agriculture and forestry sectors	9–21
Increased morbidity	6–37
Increased electricity for air conditioning	8–10
Other factors	10–18
Total	61–74

It has been estimated that, under a business-as-usual scenario, the total number of persons displaced by the impacts of global warming could total in the order of 150 million by the year 2050 (or about 3 million per year on average) – about 100 million due to sea-level rise and coastal flooding and about 50 million due to the dislocation of agricultural production mainly due to the incidence and location of areas of drought[61]. The cost of resettling 3 million displaced person per year has been estimated at between $1,000 and $5,000 per person, giving a total of about $4 thousand million per year[62]. What the estimated cost for resettlement does not include, however, (as the authors of the study themselves emphasize) is the human cost associated with displacement. Nor does it include the social and political instabilities which ensue when substantial populations are seriously disrupted because their means of livelihood has disappeared. The effects of these could be very large indeed.

The overall impact of global warming

The incidence of various impacts of global warming is complex and far from uniform over the world.

- There are many ways in which our current environment is being degraded due to human activities; global warming will tend to exacerbate these degradations. Sea-level rise will make the situation worse for low-lying land which is subsiding because of the withdrawal of groundwater and because the amount of sediment required to maintain the level of the land has been reduced. The loss of soil due to overuse of land or deforestation will be accelerated, with increasing droughts or floods in some areas. In other places, extensive deforestation will lead to drier climates and less sustainable agriculture.
- To respond to the impacts from the changes brought about by global warming, it will be necessary to adapt. In many cases this will involve changes in infrastructure, for instance new sea defences or water supplies. Many of the impacts of climate change will be adverse, but even when the impacts in the long term turn out to be beneficial, in the short term the process of adaptation will mostly have a negative impact and involve cost.
- The most important impact is on water supplies, which are in any case becoming increasingly critical in many places. Some parts of the world are expected to become warmer and drier, especially in summer, with a greater likelihood of droughts; in other parts a greater incidence of floods is expected.
- Through adaptation to different crops and practices, first indications are that the total of world food production will not be seriously affected by climate change – although studies have not yet taken into account the likely occurrence of climate extremes. However, the disparity in per capita food supplies between the developed and the developing world will almost certainly become larger.
- Because of the likely rate of climate change, there will also be a serious impact on natural ecosystems, especially at mid to high latitudes. Forests especially will be affected by increased climate stress causing substantial die-back and loss of production, associated with which is likely to be the positive feedback of additional carbon dioxide emissions. In a warmer world longer periods of heat stress will have an effect on human health; warmer temperatures will also encourage the spread of certain tropical diseases, such as malaria, to new areas.
- Economists have attempted to estimate the average annual cost in money terms of the impacts which would arise under the climate change likely for the IPCC business-as-usual scenario of greenhouse gas emissions. Averaged over the world for a time around the middle of next century, these estimates are typically around 1 to 1.5 per cent of GDP for developed countries

and around 5 per cent for developing countries. But they do not take into account the cost in human terms nor the substantial social and political disruption the impacts will bring. In particular, it is estimated that there could be up to 3 million new environmental refugees each year or over 150 million by the middle of next century.

It is important to bear in mind that these estimates have concentrated on the doubled carbon dioxide scenario (in other words, the next 50 or 60 years). Soon after the end of next century, under the IPCC business-as-usual scenario (in other words, if strong action is not taken to curb carbon dioxide emissions) a further doubling of the equivalent carbon dioxide concentration will have occurred and it will be continuing to rise. The impacts of the additional climate change which would occur with a second effective doubling of carbon dioxide are likely to be substantially more severe than those of the first doubling.

That, of course, is a lot further away in time; perhaps for that reason it has not been given much attention. However, because of the long life-time of some greenhouse gases, because of the long memory of the climate system, because some of the impacts may turn out to be irreversible and also because of the time taken for human activities and ecosystems to respond and change course, it is important to have an eye on the longer term. The much more severe impacts which can be expected at longer time horizons increase the imperative now to take the necessary action.

However, many will ask why we should be concerned about the state of the Earth so far ahead in the future. Can we not leave it to be looked after by future generations? The next chapter will give something of my personal motivation for caring about what happens to the Earth in the future as well as now.

Questions   **1**   For your local region, find out about its water supply and how the water is used (eg by domestic users, agriculture, industry etc). What are likely to be the trends in its use over the next fifty years due, for instance, to population changes or changes in agriculture or industry? What are the possibilities for increased supply and how might these be affected by climate change?

2   For your local area, find out about current environmental problems such as sea-level rise due to subsidence, over-use of ground water, air pollution affecting forests. Which of these are likely to be exacerbated by climate change? Try to estimate by how much.

3   For your local region, identify the possible impacts of climate change over the next hundred years and quantify them as far as you can. Attempt to make an estimate of the cost of the damage for each impact. How far could adaptation reduce each type of damage?

4   From the information in Chapter 6, make estimates of possible climate change by the middle of next century for typical regions of boreal forest. Then estimate from Fig. 7.11 for the each of the three tree species what loss of productivity might occur in each case.

5   Make an estimate of the total volume of ice in the Greenland and Antarctic ice-caps. What proportion would have to melt to increase the sea level by the 6 m or so which occurred during the last interglacial period?

6 In the past human communities have adapted to changes of many kinds including some changes in climate. It is sometimes argued that, because the adaptability of human beings is not fully allowed for, the likely damage from the impacts of climate change in the future tends to be overestimated. Do you agree?

Notes

1 For comprehensive detail about climate change impacts see *Climate Change 1995: Impacts, Adaptations and Mitigation of Climate Change*, eds. R. T. Watson, M. C. Zinyowera and R. H. Moss, CUP, 1996.

2 See for instance A. Goudie, *The Human Impact on the Natural Environment*, MIT Press, 1986.

3 See for instance C. J. van der Veen, 'State of balance of the cryosphere', *Rev. Geophys.*, **29**, 1991, pp. 433–55.

4 From R. A. Warrick *et al.* 'Changes in sea level', chapter 7 in Climate Change 1995 The Science of Climate Change, eds. J. T. Houghton, L. G. Meira Filho, B. A. Callender, N. Harris, A. Kattenberg and K. Maskell, CUP, 1996.

5 From R. A. Warrick and J. Oerlemans, chapter 9 in *Climate Change, the IPCC Scientific Assessment*, eds. J. T. Houghton, G. J. Jenkins and J. J. Ephraums, CUP, 1990.

6 J. M. Broadus, 'Possible impacts of, and adjustments to, sea-level rise: the cases of Bangladesh and Egypt' in *Climate and Sea-level Change: Observations, Projections and Implications*, eds. R. A. Warrick, E. M. Barrow and T. M. L. Wigley, CUP, 1993, pp. 263–75. Note that, because of variations in the ocean structure, sea-level rise would not be the same everywhere. In Bangladesh it would be somewhat above average (see J. M. Gregory 'Sea-level changes under increasing CO_2 in a transient coupled ocean-atmosphere experiment,' *J. Climate*, **6**, 1993, pp. 2247–62).

7 From J. M. Broadus, 'Possible impacts of, and adjustments to, sea-level rise: the cases of Bangladesh and Egypt' in *Climate and Sea-level Change: Observations, Projections and Implications*, eds. R. A. Warrick, E. M. Barrow and T. M. L. Wigley, CUP, 1993, pp. 263–75 – adapted from J. D. Milliman *et al.*, 'Environmental and economic implications of rising sea-level and subsiding deltas: the Nile and Bangladesh examples', *Ambio*, **18**, 1989, pp. 340–5.

8 From a report prepared in 1989 by F. U. Muhtab, *Effect of climate change and sea level rise on Bangladesh*, Commonwealth Secretariat, London.

9 J. M. Broadus, 'Possible impacts of, and adjustments to, sea-level rise: the cases of Bangladesh and Egypt' in *Climate and Sea-level Change: Observations, Projections and Implications*, eds. R. A. Warrick, E. M. Barrow and T. M. L. Wigley, CUP, 1993, pp. 263–75.

10 J. D. Milliman *et al.*, 'Environmental and economic implications of rising sea-level and subsiding deltas: the Nile and Bangladesh examples', *Ambio*, **18**, 1989, pp. 340–5.

11 From report entitled *Climate change due to the greenhouse effect and its implications for China* published in 1992 by the Worldwide Fund for *Nature*, Gland, Switzerland.

12 J. W. Day *et al.*, 'Impacts of sea-level rise on coastal systems with special emphasis on the Mississippi river deltaic plain' in *Climate and Sea-level Change: Observations, Projections and Implications*, eds. R. A. Warrick, E. M. Barrow and T. M. L. Wigley, CUP, 1993, pp. 276–96.

13 K. M. Clayton, 'Adjustment to greenhouse gas induced sea-level rise on the Norfolk coast – a case study' in *Climate and Sea-level Change: Observations, Projections and Implications*, eds. R. A. Warrick, E. M. Barrow and T. M. L. Wigley, CUP, 1993, pp. 310–21.

14 J. G. de Ronde, 'What will happen to the Netherlands if sea-level rise accelerates?' in *Climate and Sea-level Change: Observations, Projections and Implications*, eds. R. A. Warrick, E. M. Barrow and T. M. L. Wigley, CUP, 1993, pp. 322–35.

15 L. Bijlsma, 'Coastal zones and small islands', chapter 9 in *Climate Change 1995: Impacts, Adaptations and Mitigation of Climate Change*, eds. R. T. Watson, M. C. Zinyowera and R. H. Moss, CUP, 1996.

16 L. Bijlsma, 'Coastal zones and small islands', chapter 9 in *Climate Change 1995: Impacts, Adaptations and Mitigation of Climate Change*, eds. R. T. Watson, M. C. Zinyowera and R. H. Moss, CUP, 1996.

17 From J. W. Maurits la Riviere, 'Threats to the world's water', *Sci Amer.* **261**, September 1989, pp. 48–55.

18 From *The World Environment 1972–1992*, eds. M. K. Tolba and O. A. El-Kholy, Chapman and Hall, 1992, p. 84.

19 From *The World Environment 1972–1992*, eds. M. K. Tolba and O. A. El-Kholy, Chapman and Hall, 1992, p. 85.

20 Quoted by Geoffrey Lean in 'Troubled Waters', in the colour supplement to the *Observer* newspaper, 4 July 1993.

21 P. H. Gleick 'Regional hydrologic consequences of increases in atmospheric CO_2 and other trace gases', Climatic Change, 10, 1987, pp. 137–61.

22 J. A. Dracup and D. R. Kendall, 'Floods and droughts' in *Climate Change and US Water Resources*, ed. P. E. Waggoner, Wiley, 1990, pp. 243–67.

23 P. H. Gleick, 'Vulnerability of water systems' in *Climate Change and US Water Resources*, ed. P. E. Waggoner, Wiley, 1990, pp. 223–40.

24 'Towards an integrated impact assessment of climate change: the MINK study', ed. N. J. Rosenberg, *Climatic Change*, **24**, Nos 1–2, 1993, Kluwer Academic Publishers.

25 The MINK study on water resources is described by K. D. Frederick, *Climatic Change*, **24**, 1993, pp. 83–115.

26 J. Lean and P. R. Rowntree, 'A simulation of the impact of Amazonian deforestation on climate using an improved canopy representation' *Q. J. R. Meteorol. Soc.*, **119**, 1993, pp. 509–30. Similar results with somewhat larger reductions in rainfall have been reported by A. Henderson-Sellers *et al.*, 'Tropical deforestation: modelling local to regional scale climate change', *J. Geophys. Res.*, **98**, 1993, pp. 7289–315.

27 M. F. Mylne and P. R. Rowntree, 'Modelling the effects of albedo change associated with tropical deforestation', *Climatic Change*, **21**, 1992, pp. 317–43.

28 Quoted in J. F. B. Mitchell *et al.*, 'Equilibrium climate change and its implications for the future' in *Climate Change, the IPCC Scientific Assessment*, eds. J. T. Houghton, G. J. Jenkins and J. J. Ephraums, CUP, 1990, pp. 131–72.

29 See for instance J. W. Maurits la Riviere, 'Threats to the world's water', *Scien. Amer.* 261, September 1989, pp. 48–55; also P. Bullock, 'Land degradation and desertification', chapter 4 in *Climate Change 1995: Impacts, Adaptations and Mitigation of Climate Change*, eds. R. T. Watson, M. C. Zinyowera and R. H. Moss, CUP, 1996.

30 *Climate Change and US Water Resources*, ed. P. E. Waggoner, Wiley, 1990.

31 From *The World Environment 1972–1992*, eds. M. K. Tolba and O. A. El-Kholy, Chapman and Hall, 1992, p. 134.

32 From *The World Environment 1972–1992*, eds. M. K. Tolba and O. A. El-Kholy, Chapman and Hall, 1992, p. 135.

33 M. Lal, 'Water resources of the south-east Asian region in a warmer atmosphere', *Advances in Atmospheric Sciences*, **11**, No 2, 1994, Academia Sinica, China.

34 M. Lal, 'Global climate change and its impact on Indian agriculture' in a state-of-the-art publication of ICAR, New Delhi, entitled *Global Climate Change and Indian Agriculture*, 1994.

35 M. Lal *et al.*, 'Effect of transient increase in greenhouse gases and sulphate aerosols on monsoon climate', *Current Science*, **69**, 1995, pp. 752–63.

36 M. Lal, 'Water resources of the south-east Asian region in a warmer atmosphere', *Advances in Atmospheric Sciences*, **11**, No 2, 1994, Academia Sinica, China.

37 P. R. Crosson and N. J. Rosenberg, 'Strategies for agriculture', *Scien. Amer.*, 261, September 1989, pp. 78–85.

38 Information in proposal, edited by A. D. Moura, for an International Research Institute for Climate Prediction report prepared in 1992 for the International Board for the TOGA project, World Meteorological Organization, Geneva.

39 C. Rosenzweig, M. L. Parry, G. Fischer and K. Frohberg, *Climate Change and World Food Supply, Research Report No 3*, 1993, Environmental Change Unit, University of Oxford; also, J. Reilly *et al.*, 'Agriculture in a changing climate', chapter 13 in *Climate Change 1995: Impacts, Adaptations and Mitigation of Climate Change*, eds. R. T. Watson, M. C. Zinyowera and R. H. Moss, CUP, 1996.

40 For an estimate of the global effect of CO_2 fertilization on ecosystems see J. M. Melillo *et al.*, 'Global climate change and terrestrial net primary production', *Nature*, **363**, 1993, pp. 234–40.

41 *The World Environment 1972–1992*, eds. M. K. Tolba and O. A. El-Kholy, Chapman and Hall, 1992.

42 C. Rosenzweig, M. L. Parry, G. Fischer and K. Frohberg, *Climate Change and World Food Supply, Research Report No 3*, 1993, Environmental Change Unit, University of Oxford.

43 U. F. Miko *et al.*, 'Climate change impacts on forests', chapter 1 in *Climate Change 1995: Impacts, Adaptations and Mitigation of Climate Change*, eds. R. T. Watson, M. C. Zinyowera and R. H. Moss, CUP, 1996.

44 Adapted from D. M. Gates , *Climate Change and its Biological Consequences*, Sinauer Associates Inc. Sunderland, Mass., USA, 1993, p. 63.; the original source is P. A. Delcourt and H. R. Delcourt, in *Geobotany II*, ed. R. C. Romans, Plenum Press, New York, 1981, pp. 123–65. Gates' book contains a detailed review of natural ecosystems and climate change.

45 D. M. Gates, *Climate Change and its Biological Consequences*, Sinauer Associates Inc. Sunderland, Mass., USA, 1993, p. 63.

46 D. M. Gates, *Climate Change and its Biological Consequences*, Sinauer Associates Inc. Sunderland, Mass., USA, 1993, p. 77.

47 M. D. Bowes and R. A. Sedjo, 'Impacts and responses to climate change in forests of the MINK region', *Climatic Change*, **24**, 1993, pp. 63–82.

48 Data from H. Bugmann quoted in U. F. Miko *et al.*, 'Climate change impacts on forests', chapter 1 in *Climate Change 1995: Impacts, Adaptations and Mitigation of Climate Change*, eds. R. T. Watson, M. C. Zinyowera and R. H. Moss, CUP, 1996.

49 J. M. Melillo *et al.*, 'Terrestrial biotic responses to environmental change and feedbacks to climate', chapter 9 in *Climate Change 1995: The Science of Climate Change*, eds. J. T. Houghton, L. G. Meira Filho, B. A. Callender, N. Harris, A. Kattenberg and K. Maskell, CUP, 1996; also U. F. Miko *et al.*, 'Climate change impacts on forests', chapter 1 in *Climate Change 1995: Impacts, Adaptations and Mitigation of Climate Change*, eds. R. T. Watson, M. C. Zinyowera and R. H. Moss, CUP, 1996.

50 *Climate Change: the IPCC Impacts Assessment*, eds. W. J. McG. Tegart *et al.*, Australian Government Publishing Service, Canberra, 1990, pp. 6–20.

51 See A. J. McMichael, *Planetary Overload*, CUP, 1993; A. J. McMichael *et al.*, 'Human population health', chapter 18 in *Climate Change 1995: Impacts, Adaptations and Mitigation of Climate Change*, eds. R. T. Watson, M. C. Zinyowera and R. H. Moss, CUP, 1996. Also a series of weekly articles under the general heading of 'Climate and Health' in the *Lancet* from 23 October to 11 December 1993 addressed many of the issues.

52 I. S. Kalkstein, 'Direct impact in cities', *Lancet*, **342**, 1993, pp. 1397–9; also A. J. McMichael *Planetary Overload*, CUP, 1993.

53 Studies carried out by the US Environmental Protection Agency, quoted by W. R. Cline, *The Economics of Global Warming*, Institute for International Economics, Washington DC, 1992, pp. 116–17.

54 N. Nicholls, 'El Niño-Southern Oscillation and vector-borne disease', *Lancet*, **342**, 1993, pp. 1284–5. The El Niño cycle is described above in Chapter 5.

55 A. Haines *et al.*, 'Global Health Watch: monitoring impacts of environmental change', *Lancet*, **342**, 1993, pp. 1464–9.

56 D. W. Pearce *et al.*, 'The social costs of climate change', chapter 6 in *Climate Change 1995: Economic and Social Dimensions of Climate Change*, eds. J. Bruce, Hoesung Lee and E. Haites, CUP, 1996.

57 W. R. Cline, *The Economics of Global Warming*, Institute for International Economics, Washington DC, 1992.

58 Studies by Cline, Fankhauser and Tol: results presented in D. W. Pearce *et al.*, 'The social costs of climate change', chapter 6 in *Climate Change 1995: Economic and Social Dimensions of Climate Change*, eds. J. Bruce, Hoesung Lee and E. Haites, CUP, 1996.

59 D. W. Pearce *et al.*, 'The social costs of climate change', chapter 6 in *Climate Change 1995: Economic and Social Dimensions of Climate Change*, eds. J. Bruce, Hoesung Lee and E. Haites, CUP, 1996.

60 D. W. Pearce *et al.*, 'The social costs of climate change', chapter 6 in *Climate Change 1995: Economic and Social Dimensions of Climate Change*, eds. J. Bruce, Hoesung Lee and E. Haites, CUP, 1996.

61 N. Myers and J. Kent, 'Environmental exodus: an emergent crisis in the global arena', published by Climate Institute, Washington DC; also N. Adger and S. Fankhauser, 'Economic analysis of the greenhouse effect: optimal abatement level and strategies for mitigation', *Int. J. of Environment and Pollution*, **3**, 1993, pp. 104–19.

62 N. Adger and S. Fankhauser, 'Economic analysis of the greenhouse effect: optimal abatement level and strategies for mitigation', *Int. J. of Environment and Pollution*, **3**, 1993, pp. 104–19.

8 Why Should We Be Concerned?

I have been describing the likely changes in climate which may occur as a result of human activities, and the impact these may have in different parts of the world. But large and potentially devastating changes are likely to be a generation or more away. So why should we be concerned? What responsibility, if any, do we have for the planet as a whole and the great variety of other forms of life which inhabit it? And does our scientific knowledge in any way match up with other insights, for instance ethical and religious ones, regarding our relationship with our environment? In this chapter I want to digress from the detailed consideration of global warming (to which I shall return) in order briefly to explore these fundamental questions and to present something of my personal viewpoint on them.

Earth in the Balance

Al Gore, the Vice-President of the United States, entitled his book on the environment *Earth in the Balance*[1], implying that there are balances in the environment which need to be maintained. A small area of a tropical forest possesses an ecosystem which contains some thousands of plant and animal species, each thriving in its own ecological niche in close balance with the others. Balances are also important for larger regions and for the Earth as a whole. These balances can be highly precarious, especially where humans are concerned.

One of the first to point this out was Rachel Carson in her book *Silent Spring*[2], first published in 1962, which described the damaging effects of pesticides on the environment. Humans are an important part of the global ecosystem; as the size and scale of human activities continue to escalate, so can the seriousness of the disturbances caused to the overall balances of nature. Some examples of this were given in the last chapter.

It is important that we recognize these balances, in particular the careful relationship between humans and the world around us. It needs to be a balanced and harmonious relationship in which each generation of humans should leave the Earth in a better state, or at least in as good a state as they found it. The word that is often used for this is sustainability – politicians talk of sustainable development (see box in Chapter 9 p. 163). This principle, and its link with the harmonious relationship between humans and nature, was given prominent place by the United Nations Conference on Environment and Development held at Rio de Janeiro in Brazil in June 1992. The first principle in a list of 27 at the Rio Declaration adopted by the Conference is 'Human beings are at the centre of concerns for sustainable development. They are entitled to a healthy and productive life in harmony with nature.'

However, despite such statements of principle from a body such as the United Nations, many of the attitudes which we commonly have to the Earth are neither balanced, harmonious nor sustainable. Some of these are briefly outlined in the following paragraphs.

Exploitation

Humankind has over many centuries been exploiting the Earth and its resources. It was at the beginning of the Industrial Revolution some two hundred years ago that the potential of the Earth's minerals began to be realized. Coal, the result of the decay of primaeval forests and laid down over many millions of years, was the main source of energy for the new industrial developments. Iron ore to make steel was mined in vastly increased quantities. The search for other metals such as zinc, copper and lead was intensified until today many millions of tons are mined each year. Around 1960, oil took over from coal as the dominant world source of energy; oil and gas between them now supply over twice the energy supplied by coal.

We have not only been exploiting the Earth's mineral resources. The Earth's biological resources have also been attacked. Forests have been cut down on a large scale to make room for agriculture and for human habitation. Tropical forests are a particularly valuable resource, important for the maintenance of the climate of tropical regions. They have also been estimated to contain perhaps half of all the Earth's biological species. Yet only about half of the mature tropical forests which existed a few hundred years ago still stand[3]. At the present rate of destruction virtually all will be gone in less than fifty years.

Much of this exploitation has been carried out with little or no thought as to whether this use of natural resources is a responsible one. Early in the Industrial Revolution it seemed that resources were essentially limitless. Later on, as one source ran out others became available to more than take its place. Even now, for most minerals new sources are being found faster than present sources are being used. But the growth of use is such that this situation cannot continue. In many cases known reserves or even likely reserves will begin to run out during the next hundred or few hundred years. These resources have been laid down over many millions if not billions of years. Nature took about a million years to lay down the amount of fossil fuel that we now burn worldwide every year – and in doing so it seems that we are causing rapid change of the Earth's climate. Such a level of exploitation is clearly not in balance, not harmonious and not sustainable.

'Back to nature'

Almost the reverse of this attitude is the suggestion that we all adopt a much more primitive lifestyle and give up a large part of industry and intensive farming – that we effectively put the clock back two or three hundred years to before the Industrial Revolution. That sounds very seductive and some individuals can clearly begin to live that way. But there are two main problems.

The first is that it is just not practical. The world population is now some six times what it was two hundred years ago and about three times that of fifty years ago. The world cannot be adequately fed without farming on a reasonably intensive scale and without modern methods of food distribution. Further, most people that have them would not be prepared to be without the technical aids – electricity, central heating, refrigerator, washing machine, television and so on – which give the freedom, the interest and the entertainment which is so much taken for granted. Moreover, increasing numbers of people in the developing world are also taking advantage of and enjoy these aids to a life of less drudgery and more freedom.

The second problem is that it fails to take account of human creativity. Human scientific and technical development cannot be frozen at a given point in history, insisting that no further ideas can be developed. A proper balance between humans and the environment must leave room for humans to exercise their creative skills.

Again, therefore, a 'back to nature' viewpoint is neither balanced nor sustainable.

The technical fix

A third common attitude to the Earth is to invoke the 'technical fix'. As a senior environmental official from the United States said to me some years ago, 'We cannot change our lifestyle because of the possibility of climate change, we just need to fix the biosphere.' It was not clear just what he supposed the technical fixes would turn out to be. The point that he was making is that, in the past, humans have been so effective at developing new technology to meet the problems as they arise, can it not be assumed that this will continue? Concern about the future then turns into finding the 'fixes' as they are required.

On the surface the 'technical fix' route may sound a good way to proceed; it demands little effort and no foresight. It implies that damage can be corrected when it has been created rather than avoided in the first place. But damage already done to the environment by human activities is causing problems now. It is as if in looking after my home I decided not to carry out any routine maintenance but 'fixed' the failures as they occurred. For my home that would be a high risk route to follow: failure to rewire when necessary could easily lead to a disastrous fire. A similar attitude to the Earth is both arrogant and irresponsible. It fails to recognize the vulnerability of nature to the large changes which human activities are now able to generate.

Science and technology possess enormous potential to assist in caring for the Earth, but they must be employed in a careful, balanced and responsible way. The 'technical fix' approach is neither balanced nor sustainable.

Future generations

Having described attitudes which are not balanced or harmonious in their relationship to the Earth and which fail to contribute to sustainability, I now turn to describe attitudes to the environment which are more acceptable in terms of the criteria I have set.

Firstly, there is our responsibility to future generations. It is a basic instinct that we wish to see our children and our grandchildren well set up in the world and wish to pass on to them some of our most treasured possessions. A similar desire would be that they inherit from us an Earth which has been well looked after and which does not pose to them more difficult problems than those we have had to face. But such an attitude is not universally held. I remember well, after a presentation I made on global warming to the British Cabinet at number 10, Downing Street in London, a senior politician commented that the problem would not become serious in his lifetime and could be left for its solution to the next generation. I do not think he had appreciated that the longer we delay in taking action, the larger the problem becomes and the more difficult to solve. There is a need to face up to the problem now for the sake of the next and subsequent generations. We have no right to act as if there is no tomorrow. We also have a responsibility to give to those who follow us a pattern for their future based on the principle of sustainable development.

The unity of the Earth

A second point of view sees us as having some responsibility, not just for all generations of humanity, but also for the larger world of all living things. We are, after all, part of that larger world. There is good scientific justification for this. We are becoming increasingly aware of our dependence on the rest of nature and of the interdependencies which exist between different forms of life, between living systems and the physical and chemical environment which surrounds life on the Earth – and indeed between ourselves and the rest of the universe.

The scientific theory named Gaia after the Greek Earth goddess and publicized particularly by James Lovelock emphasizes these interdependencies. Lovelock[4] points out that the chemical composition of the Earth's atmosphere is very different from that of our nearest planetary neighbours, Mars and Venus. Their atmospheres, apart from some water vapour, are almost pure carbon dioxide. The Earth's atmosphere, by contrast, is 78 per cent nitrogen, 21 per cent oxygen and only 0.03 per cent carbon dioxide. So far as the major constituents are concerned, this composition has remained substantially unchanged over many millions of years – a fact that is very surprising when it is realized that it is a composition that is very far from chemical equilibrium.

This very different atmosphere on the Earth has come about because of the emergence of life. Early in the history of life, plants appeared which photosynthesize, taking in carbon dioxide and giving out oxygen. There followed other living systems which 'breathe', taking in oxygen and giving out carbon dioxide. The presence of life therefore influences and effectively controls the environment to which living systems in turn adapt. It is the close match of the environment to the needs of life and its development which seems so remarkable and which Lovelock has emphasized. He gives many examples; I will quote one concerned with oxygen in the atmosphere. There is a critical connection between the oxygen concentration and the frequency of forest fires[5]. Below an oxygen concentration of 15 per cent, fires cannot be started even in dry twigs. At concentrations above 25 per cent fires burn extremely fiercely even in the damp wood of a tropical rain forest. Some species are dependent on fires for their survival; for instance, some conifers require the heat of fire to release their seeds from the seed pods. Above 25 per cent concentration of oxygen there would be no forests; below 15 per cent, the regeneration that fires provide in the world's forests would be absent. The oxygen concentration of 21 per cent is ideal.

It is this sort of connection that has driven Lovelock to propose that there is tight coupling between the organisms that make up the world of living systems and their environment. He has suggested a simple model of an imaginary world called Daisyworld (see box) which illustrates the type of feedback mechanisms which can lead to tight coupling and exert control. This model is similar to the one he has proposed for the biological and chemical history of the Earth during the first 1,000 million years after primitive life first appeared on the Earth some 3,500 million years ago.

The real world is, of course, enormously more complex than Daisyworld, which is why the Gaia hypothesis has led to so much debate. Lovelock's first statement in 1972 of the hypothesis[6] was that 'Life, or the biosphere, regulates or maintains the climate and the atmospheric composition at an optimum for itself.' In his later writings he has introduced the analogy between the Earth and a living organism, introducing a new science which he calls geophysiology[7] – a more recent book is entitled *Gaia, the Practical Science of Planetary Medicine*[8].

An advanced organism such as a human being has many built-in mechanisms for controlling the interactions between different parts of the organism and for self-regulation. In a similar way, Lovelock argues, the ecosystems on the Earth are so tightly coupled to their physical and chemical environments that the ecosystems and their environment could be considered as one organism with an integrated 'physiology'. In this sense he believes that the Earth is 'alive'.

That elaborate feedback mechanisms exist in nature for control and for adaptation to the environment is not in dispute. But many scientists feel that Lovelock has gone too far in suggesting that ecosystems and their environment can be considered as a single organism. Although Gaia has stimulated much scientific comment it remains a hypothesis[9]. What the debate has done, however, is to emphasize the interdependencies which connect all living systems to their environment – the biosphere is a system in which is incorporated a large measure of self-control.

There is the hint of a suggestion in the Gaia hypothesis that the Earth's feedbacks and self-regulation are so strong that we humans need not be concerned about the pollution we produce – Gaia has enough control to take care of anything we might do. Such a view fails to recognize the effect on the Earth's system of substantial disturbances, in particular the vulnerability of the environment with respect to its suitability for humans. To quote Lovelock[10], 'Gaia, as I see her, is no doting mother tolerant of misdemeanours, nor is she some fragile and delicate damsel in danger from brutal mankind. She is stern and

Daisyworld and life on the early Earth

Daisyworld is an imaginary planet spinning on its axis and orbiting a sun rather like our own. Only daisies live in Daisyworld; they are of two hues, black and white. The daisies are sensitive to temperature. They grow best at 20 °C, below 5 °C they will not grow and above 40 °C they wilt and die. The daisies influence their own temperature by the way they absorb and emit radiation; black ones absorb more sunlight and therefore keep warmer than white ones.

In the early period of Daisyworld's history (Fig. 8.1), the sun is relatively cool and the black daisies are favoured because, by absorbing sunlight, they can keep their temperature closest to 20 °C. Most of their white cousins die because they reflect sunlight and fail to keep above the critical 5 °C. However, later in the planet's history, the sun becomes hotter. Now the white daisies can also flourish; both sorts of daisies are present in abundance. Later still as the sun becomes even hotter the white daisies become dominant as conditions become too warm for the black ones. Eventually, if the sun continues to increase its temperature even the white ones cannot keep below the critical 40 °C and all the daisies die.

Daisyworld is a simple model employed by Lovelock[11] to illustrate the sort of feedbacks and self-regulation which occur in very much more complex forms within the living systems on the Earth.

Lovelock proposes a similar simple model as a possible description of the early history of life on the Earth (Fig. 8.2). The dashed line shows the temperature which would be expected on a planet possessing no life but with an atmosphere consisting, like our present atmosphere, mostly of nitrogen with about 10 per cent carbon dioxide. The rise in temperature occurs because the sun gradually became hotter during this period. About 3,500 million years ago primitive life appeared. Lovelock, in this model, assumes just two forms of life, bacteria which are anaerobic photosynthesizers – using carbon dioxide to build up their bodies but not giving out oxygen – and bacteria which are decomposers, converting organic matter back to carbon dioxide and methane. As life appears the temperature decreases as the concentration of the greenhouse gas, carbon dioxide, decreases. At the end of the period about 2,300 million years ago, more complicated life appears; there is an excess of free oxygen and the methane abundance falls to low values, leading to another fall in temperature, methane also being a greenhouse gas. The overall influence of these biological processes has been to maintain a stable and favourable temperature for life on the Earth.

Fig. 8.1
Daisyworld.

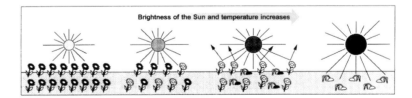

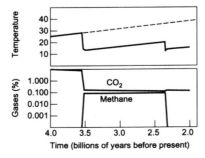

Fig. 8.2.
Model of the Earth's early history, as proposed by Lovelock[12].

tough, always keeping the world warm and comfortable for those who obey the rules, but ruthless in her destruction of those who transgress. Her unconscious goal is a planet fit for life. If humans stand in the way of this, we shall be eliminated with as little pity as would be shown by the micro-brain of an intercontinental ballistic nuclear missile in full flight to its target.'

The Gaia scientific hypothesis can help to bring us back to recognize two things, firstly the inherent value of all parts of nature and secondly our dependence, as human beings, on the Earth and on our environment. Gaia is a scientific theory. But some have been quick to see it as a religious idea, supporting ancient religious beliefs. Many of the world's religions have drawn attention to the close relationship between humans and the Earth.

The Native American tribes of North America lived close to the Earth. One of their chiefs when asked to sell his land expressed his dismay at the idea and said[13], 'The Earth does not belong to man, man belongs to the Earth. All things are connected like the blood that unites us all.' An ancient Hindu saying[14], 'The Earth is our mother, and we are all her children' also emphasizes a feeling of closeness to the Earth. Those who have worked closely with indigenous peoples have given many examples of the care with which, in a balanced way, they look after the trees, plants and animals in their local ecosystem[15].

The Islamic religion teaches the value of the whole environment, for instance in a saying of the prophet Mohammed: 'He who revives a dead land will be rewarded accordingly, and that which is eaten by birds, insects and animals out of that land will be charity provided by God' – so emphasizing both our duty to care for the natural environment and our obligation to allow all living creatures their rightful place within it[16].

Judaism and Christianity share the stories of creation in the early chapters of the Bible which emphasize the responsibility of humans to care for the Earth – we shall refer to these stories again later on in the chapter. Further on in the Old Testament are passages which give detailed instructions regarding care for the land and the environment[17]. Christianity has been described by William Temple, Archbishop of Canterbury sixty years ago, as 'the most materialistic of the great religions'. Because of its central belief that God became human in Jesus (an event Christians call the incarnation), Temple goes on to say 'by the very nature of its central doctrine Christianity is committted to a belief...in the reality of matter and its place in the divine scheme[18]'. For the Christian, the twin doctrines of creation and incarnation demonstrate God's interest in and concern for the Earth and the life it contains.

In looking for themes which emphasize the unity between humans and their environment, we need not confine ourselves to the Earth. There is a very much larger sphere in which a similar perspective of unity is becoming apparent. Some astronomers and cosmologists, overwhelmed by the size, scale, complexity, intricacy and precision of the universe, have begun to realize that their quest for an understanding of the evolution of the universe right from the 'Big Bang' some fifteen thousand million years ago is not just a scientific project but a search for meaning[19]. Why else has Stephen Hawking's book *A Brief History of Time*[20], in selling over six million copies, become one of the bestsellers of our time?

In this new search for meaning, the perspective has arisen that the universe was made with humans in mind – an idea expressed in some formulations of the 'anthropic principle'[21]. Two particular pointers emphasize this. Firstly, we have already seen that the Earth itself is fitted in a remarkable way for advanced forms of life. Cosmology is telling us that, in order for life on our planet to be possible, the universe itself at the time of the Big Bang and in its early history needed to be 'fine-tuned' to an incredible degree[22]. Secondly, there is the remarkable fact that human minds, themselves dependent on the whole universe for their existence, are able to appreciate and understand to some extent the fundamental mathematical structure of the universe's design[23]. As Albert Einstein commented, 'The most incomprehensible thing about the universe is that it is comprehensible.' In the theory of Gaia, the Earth itself is central and humans are just one part of life on Earth; the insights of cosmology suggest that humans have a particular place in the whole scheme of things.

This section has recognized the intrinsic unity and interdependencies which exist not only on our Earth but also within the whole universe, and the particular place that we humans have in the universe. Being aware of these has large implications for our attitude to our environment.

Environmental values

What do we value in the environment and how do we decide what we need to preserve, to foster or improve? At the basis of our discussion so far have been several assumptions regarding the value or importance of different fundamental attitudes or actions, some of which I have associated with ideas which come from the underlying environmental science. Is it legitimate, however, to make connections of this kind between science and values? It is often argued that science itself is value free. But science is not an activity in isolation. As Polanyi[24] has pointed out, the facts of science cannot sensibly be considered apart from the participation and the commitment of those who discover those facts or incorporate them into wider knowledge.

In the methodology and the practice of science are many assumptions of value. For instance, that there is an objective world of value out there to discover, that there is value in the qualities of elegance and economy in scientific theory, that complete honesty and cooperation between scientists are essential to the scientific enterprise.

Values can also be suggested from the perspective of the underlying science as we have shown earlier in the chapter. For instance, we have described the Earth in terms of balance, interdependency and unity. Since all of these are critical to the Earth as we know it, we can argue that they are of fundamental value and worth preserving. We have also provided some scientific evidence that humans have a particular place in the overall scheme of the natural world, that they possess special knowledge – which suggests that they also possess special responsibility.

Moving away from science, we have already referred to values related to the environment which come from our basic experiences as human beings. These are often called 'shared values' because they are common to different members of a human community – which may be a local community, a nation or ultimately the global community taking in the whole human race. An outstanding example is the conservation of the Earth and its resources, not just for our generation but for future generations. Other examples may involve how resources are used now for the benefit of the present generation of humans and how they are shared between different communities or nations. When shared values are applied to real situations, however, conflicts often arise. For instance, how much should we forego now in order to make provision for future generations, or how should resources be shared between different countries, for instance between those in the relatively rich 'North' and those in the relatively poor 'South'? Discussion within and between human communities can assist in the definition and application of such shared values.

Many of these shared values have their origins in the cultural and religious backgrounds of human communities. Discussions about values need therefore to recognize fully the cultural and religious traditions, beliefs and assumptions which underly many of our attitudes and reasoning about ethical concerns.

An obstacle to the recognition of religious assumptions in the attempt to establish environmental values is the view that religious belief is not consistent with a scientific outlook. Some scientists maintain that only science can provide real explanations based on provable evidence whereas the assertions of religion cannot be tested in an objective way[25]. Other scientists, however, have suggested that the seeming inconsistency between science and religion arises because of misunderstandings about the questions being addressed by the two disciplines and that there is more in common between the methodologies of science and religion than is commonly thought[26].

Scientists are looking for descriptions of the world which fit in to an overall scientific picture. They are working towards making this picture as complete as possible. For instance, scientists are looking for mechanisms to describe the 'fine-tuning' of the universe (these are known as 'Theories of Everything'!) mentioned earlier. They are also looking for mechanisms to describe the interdependencies between living systems and the environment.

But the scientific picture can only depict part of what concerns us as human beings. Science deals with questions of 'how' not questions of 'why'. Most questions about values are 'why' questions. Nevertheless, scientists do not always draw clear distinctions between the two. Their motivations have often been associated with the 'why' questions. That was certainly true of the early scientists in the sixteenth and seventeenth centuries many of whom were deeply religious and whose main driving force in pursuit of the new science was that they might 'explore the works of God'[27].

That science and religion should be seen as complementary ways of looking at truth is a point made strongly by Al Gore in *Earth in the Balance*[28] which lucidly discusses current environmental issues such as global warming. He blames much of our lack of understanding of the environment on the modern approach which tends to separate scientific study from religious and ethical issues. Science and technology are often pursued with a clinical detachment and without thinking about the ethical consequences. 'The new power derived from scientific knowledge could be used to dominate nature with moral impunity[29],' he writes. He goes on to describe the modern technocrat as 'this barren spirit, precinct of the disembodied intellect, which knows the way things work but not the way they are'[30]. However, he also points out[31] that 'there is now a powerful impulse in some parts of the scientific community to heal the breach' between science and religion. In particular, as we pursue an understanding of the Earth's environment, it is essential that scientific studies and technological inventions are not divorced from their ethical and religious context.

Stewards of the Earth

The relationship between humans and the Earth which I have been advocating is often described as one of stewardship. We are on the Earth as its stewards. The word implies that we are carrying out our duty as stewards on behalf of someone else – but whom? Some environmentalists see no need to answer the question specifically, others might say we are stewards on behalf of future generations or on behalf of a generalized humanity. A religious person would want to be more specific and say that we are stewards on behalf of God. The religious person would also argue that to associate the relationship of humans to God with the relationship of humans to the environment is to place the latter relationship in a wider more integrated context – providing additional insights and a more complete basis for environmental stewardship[32].

In the Judaeo-Christian tradition in the story of creation in the early chapters of the Bible is a helpful 'model' of stewardship – that of humans being 'gardeners' of the Earth. It is not only appropriate for those from those particular traditions – it is a model which can be widely applied. That story tells that humans were created to care for the rest of creation – the idea of human stewardship of creation is a very old one – and were placed in a garden, the Garden of Eden, 'to work it and take care of it'[33]. The animals, birds and other living creatures were brought to Adam in the garden for him to name them[34]. We are left with a picture of the first humans as 'gardeners' of the Earth – what does our work as 'gardeners' imply? I want to suggest four things.

- A garden provides food and water and other materials to sustain life and human industry. Part of the garden in the Genesis story contained mineral resources – 'the gold of that land is good; aromatic resin and onyx are also there'[35]. The Earth provides resources of many kinds for humans to use as they are needed.
- A garden is to be maintained as a place of beauty. The trees in the Garden of Eden were 'pleasing to the eye'[36]. Humans are to live in harmony with the rest of creation and to appreciate the value of all parts of creation. Indeed, a garden is a place where care is taken to preserve the multiplicity of species, in particular those that are most vulnerable. Millions of people each year visit gardens which have been especially designed to show off the incredible variety and beauty of nature. Gardens are meant to be enjoyed.
- A garden is a place where humans, created as described in the Genesis story in the image of God[37], can themselves be creative. Its resources provide for great potential. The variety of species and landscape can be employed to increase the garden's beauty and its productivity. Humans have learnt to generate new plant varieties in abundance and to use their scientific and

technological knowledge coupled with the enormous variety of the Earth's resources to create new possibilities for life and its enjoyment. However, the potential of this creativity is such that increasingly we need to be aware of where it can take us; it has potential for evil as well as for good. Further, good gardeners intervene in natural processes with a good deal of restraint.

- A garden is to be kept so as to be of benefit to future generations. In this context, I shall always remember Gordon Dobson, a distinguished scientist, who in the 1920s developed new means for the measurement of ozone in the atmosphere. His home outside Oxford in England possessed a large garden with many fruit trees. When he was 85, a year or so before he died, I remember finding him hard at work in his garden replacing a number of apple trees; in doing so he clearly had future generations in mind.

How well do we humans match up to the description of ourselves as gardeners caring for the Earth? Not very well, it must be said; we are more often exploiters and spoilers than cultivators. Some blame science and technology for the problems, although the fault must lie with the craftsman rather than with the tools! Others have tried to place part of the blame on attitudes[38] which they believe originate in the early chapters of Genesis, which talk of human beings having rule over creation and subduing it[39]. Those words, however, should not be taken out of context – they are not a mandate for unrestrained exploitation. The Genesis chapters also insist that human rule over creation is to be exercised under God, the ultimate ruler of creation, and with the sort of care exemplified by the picture of humans as 'gardeners'. Why, therefore do humans so often fail to get their act together?

The will to act

Many of the principles I have been enunciating are included at least implicitly in the declarations, conventions and resolutions which came out of the United Nations Conference on Environment and Development held in Rio de Janeiro in June 1992; indeed, they form the background of many statements emanating from the United Nations or from official national sources. We are not short of statements of ideals. What tend to be lacking are the capability and resolve to carry them out. Sir Crispin Tickell, a British diplomat who has lectured widely on the policy implications of climate change, has commented 'Mostly we know what to do but we lack the will to do it'[40].

Many recognize this lack of will to act as a 'spiritual' problem (using the word spiritual in a general sense), meaning that we are too obsessed with the 'material' and the immediate and fail to act according to generally accepted values and ideals particularly if it means some cost to ourselves or if it is concerned with the future rather than with the present. We are only too aware of the strong temptations we experience at both the personal and the national levels to use the world's resources to gratify our selfishness and greed. Because of this, it has been proposed that at the basis of stewardship should be a principle extending what has traditionally been considered wrong – or in religious parlance as sin – to include unwarranted pollution of the environment or lack of care for it[41].

Those with religious belief tend to emphasize the importance of coupling together the relationship of humans to the environment to the relationship of humans to God[42]. It is here, religious believers would argue, that a solution for the problem of 'lack of will' can be found. That religious belief can provide an important driving force for action is often also recognized by those who look elsewhere than religion for a solution.

One of the main messages of this chapter is that action addressing environmental problems depends not only on knowledge about them but on the values we place on the environment and our attitudes towards it. In the chapter I have suggested that assessments of environmental value and appropriate attitudes can be developed from the following:

- the perspectives of balance, interdependence and unity in the natural world generated by the underlying science;
- a recognition – some would argue suggested by the science – that humans have a special place in the universe, which in turn implies that humans have special responsibilities with respect to the natural world;
- a recognition that to damage the environment or to fail to care for it is to do wrong;
- an interpretation of human responsibility in terms of stewardship of the Earth, based on an understanding of wide obligations towards all life on Earth based on 'shared' values generally recognized by different human communities.
- a recognition of the importance of the cultural and religious basis for the principles of stewardship – humans as 'gardeners' of the Earth is a possible 'model' of such stewardship;
- a recognition that, just as the totality of damage to the environment is the sum of the damage done by a large number of individuals, the totality of action to address environmental problems is the sum of a large number of individual actions. To quote from Edmund Burke, a nineteenth century British politician: 'nobody made a greater mistake than he who did nothing because he could only do a little.'

Finally, let me recall some words of Thomas Huxley, an eminent biologist from last century, who emphasized the importance in the scientific enterprise of 'humility before the facts'. An attitude of humility is also one which lies at the heart of responsible stewardship of the Earth.

In the next chapter we shall reflect on the uncertainties associated with the science of global warming and consider how they can be taken into account in addressing the imperative for action. For instance, should action be taken now or should we wait until the uncertainties are less before deciding on the right action to take?

Questions   1   There is a debate regarding the relationship of humans to the environment. Should humans be at the centre of the environment with everything else and other life related to the human centre – in other words an anthropocentric view? Or should higher prominence be given to the non-human part of nature in our scheme of things and in our consideration of values – a more ecocentric view? If so, what form should this higher prominence take?

2 How far can science be involved in the the generation and application of environmental values?

3 How far do you think can environmental values be generated through debate and discussion in a human community without reference to the cultural or religious background of that community?

4 It has been suggested that religious belief (especially strongly held belief) is a hindrance in the debate about environmental values. Do you agree?

5 Should we strive for universally accepted values with respect to the environment? Or is it acceptable for different communities to possess different values?

6 An argument for religious belief which is sometimes put forward, irrespective of whether the belief is considered to have any foundation, is that such belief motivates people more strongly than other driving forces. Do you agree with this argument?

7 Explain how the cultural or religious traditions in which you have been brought up have influenced your view of environmental concern or action. How have these influences been modified because you now hold (or do not hold) definite religious beliefs?

8 Discuss the term 'stewardship' which is often used as a description of the relation of humans to the environment. Does it imply too anthropocentric a relationship?

9 Discuss the model of humans as 'gardeners' of the Earth. How adequate is the picture it presents of the relationship of humans to the environment?

10 Do you agree with Thomas Huxley when he spoke of the importance of humility before the scientific facts? How important do you think is humility in this context and in the wider context of the application of scientific knowledge to environmental concern?

11 Because of the formidability of the task of stewardship of the Earth, some have suggested that it is beyond the capability of the human race to tackle it adequately. Do you agree?

Notes

1 Al Gore, *Earth in the Balance*, Houghton Mifflin Company, 1992.
2 Rachel Carson, *Silent Spring*, Houghton Mifflin Company, 1962.
3 For more information see G. Lean, D. Hinrichsen, A. Markham, *Atlas of the Environment*, Arrow Books, 1990.
4 J. E. Lovelock, *Gaia*, OUP, 1979 and *The Ages of Gaia*, OUP, 1988.
5 J. E. Lovelock, *The Ages of Gaia*, OUP, 1988, pp. 131–3.
6 J. E. Lovelock and L. Margulis, *Tellus*, **26**, 1974, pp. 1–10.
7 J. E. Lovelock, 'Hands up for the Gaia hypothesis', *Nature*, **344**, 1990, pp. 100–12.
8 J. E. Lovelock, *Gaia: the practical science of planetary medicine*, Gaia Books, 1991.
9 Colin Russell in *'The Earth, Humanity and God'*, UCL Press 1994, discusses Gaia as a scientific hypothesis and also its possible religious connections.
10 J. E. Lovelock, *The Ages of Gaia*, OUP, 1988, p. 212.
11 For more details see J. E. Lovelock, *The Ages of Gaia*, OUP, 1988.

12 J. E. Lovelock, *The Ages of Gaia*, OUP, 1988, p. 82.

13 Quoted by Al Gore, *Earth in the Balance*, Houghton Mifflin Company, 1992, p. 259.

14 Quoted by Al Gore, *Earth in the Balance*, Houghton Mifflin Company, 1992, p. 261.

15 Ghillean Prance, Director of Kew Gardens in the UK, in his book *The Earth Under Threat* (Wild Goose Publications, Glasgow 1996), provides examples from his extensive work in countries of south America.

16 M. H. Khalil, 'Islam and the Ethic of Conservation', *Impact* (Newsletter of the Climate Network Africa), December 1993, p. 8.

17 A number of injunctions were given to the Jews in the Old Testament regarding care for plants and animals and care for the land, for example Leviticus 19:23–25, Leviticus 25:1–7, Deuteronomy 25:4.

18 W. Temple, *Nature, Man and God*, Macmillan, 1964 (1st edition 1934)

19 See for instance Paul Davies, *The Mind of God*, Simon and Schuster, 1992. I have also addressed this theme in J. T. Houghton, *The Search for God; can science help?* Lion Publishing, 1995.

20 Stephen Hawking, *A Brief History of Time*, Bantam Books, 1989.

21 See for instance Paul Davies, *The Mind of God*, Simon and Schuster, 1992; also J. Barrow and F. J. Tipler, *The Anthropic Cosmological Principle*, OUP, 1986.

22 J. Barrow and F. J.Tipler *The Anthropic Cosmological Principle*, OUP, 1986, and J. Gribbin and M. Rees, *Cosmic Coincidences*, Black Swan, 1991.

23 Paul Davies, *The Mind of God*, Simon and Schuster, 1992.

24 M. Polanyi, *Personal Knowledge*, London, Routledge and Kegan Paul, 1962.

25 See for instance R. Dawkins, *The Blind Watchmaker*, Longmans, 1986.

26 See for instance J. Polkinghorne, *One World*, SPCK, 1986, *Beyond Science*, CUP, 1996; J. T. Houghton, *The Search for God; can science help?*, Lion Publishing, 1995.

27 See, for instance Colin Russell *Cross-currents: Interactions between Science and Faith*, Intervarsity Press, 1985.

28 Al Gore, *Earth in the Balance*, Houghton Mifflin Company, 1992.

29 Al Gore, *Earth in the Balance*, Houghton Mifflin Company, 1992, p. 252.

30 Al Gore, *Earth in the Balance*, Houghton Mifflin Company, 1992, p. 265.

31 Al Gore, *Earth in the Balance*, Houghton Mifflin Company, 1992, p. 254.

32 For modern expositions of a Christian view of the Environment, see R. Elsdon, *Greenhouse Theology*, Monarch, 1992 and Colin Russell, *The Earth, Humanity and God*, UCL Press, London, 1994.

33 Genesis 2:15.

34 Genesis 2:19.

35 Genesis 2:12.

36 Genesis 2:9.

37 Genesis 1:27.

38 The best-known exposition of this position is L. White Jnr in, for instance, 'The historical roots of our ecological crisis', *Science*, **155**, 1987, pp. 1203–7; see Colin Russell, *The Earth, Humanity and God*, UCL Press, London, 1994, for a commentary on this thesis.

39 Genesis 1:26–28.

40 *The Doomsday Letters*, broadcast on BBC Radio 4, UK, 1996.

41 This was the first of the principles which came out of a symposium (called Patmos Principles since the climax of the symposium, held in celebration of the 1900th anniversary of the writing of the Book of Revelation, was on the island of Patmos) I attended in 1995 sponsored by the Ecumenical Patriarch Bartholomew I of the Greek Orthodox Church and Prince Philip in his capacity as President of the World

Wild Life Fund. An extremely eclectic group, scientists, politicians, environmentalists, theologians attended from a wide range of religious backgrounds and beliefs. John, the Metropolitan of Pergamon, who was chairman of the symposium's scientific committee kept emphasizing that we should consider pollution of the environment – or lack of care for the environment as a sin – not only against nature but a sin against God. His message struck a strong chord with the symposium. The principle goes on to explain that this new category of sin should include activities that lead to 'species extinction, reduction in genetic diversity, pollution of the water, land and air, habitat destruction and disruption of sustainable life styles'. The symposium's report is to be published under the title *Revelation and the Environment – AD95–1995*, (ed. Sarah Hobson and Jane Lubchenco), World Scientific Publishing, 1997.

42 In Judaeo-Christian teaching the coupling of these two relationships begins with the Creation stories in Genesis. These stories go on to describe how humans disobeyed God (chapter 3) and broke the partnership. But the Bible continually explains how God offers a way back to partnership. A few chapters on in Genesis (9:8–17), the basis of the relationship between God and Noah is a covenant agreement in which 'all life on the Earth' is included as well as humans. A relationship based on covenant is also the basis of the partnership between God and the Jewish nation in the Old Testament. But, after many times when that relationship was broken, the Old Testament prophets looked forward to a new covenant based not on law but on a real change of heart (Jeremiah 31:31–34).

The New Testament writers (for example Hebrews 8:10–11) see this new covenant being worked out through the life and particularly through the death and resurrection of Jesus, the Son of God. Jesus promised his followers the Holy Spirit (John 15, 16), whose influence would enable the partnership between them and God to work. Paul, in his letters, is constantly referring to the dependent relationship which forms the basis of his own partnership with God (Galatians 2:20, Philippians 4:13) and which has been the experience of millions of Christians down the centuries. Included in Paul's theology is the whole of creation (Romans 8:19–22).

9 Weighing the Uncertainty

This book is intended to present clearly the current scientific position on global warming. A key part of this presentation must concern the uncertainty associated with all parts of the scientific description, especially with the prediction of future climate change, which forms an essential consideration when decisions regarding action are being taken. However, uncertainty is a relative term; utter certainty is not often demanded on everyday matters as a prerequisite for action. Here the issues are complex; we need to consider how uncertainty is weighed against the cost of possible action.

The scientific uncertainty

In earlier chapters I explained in some detail the science underlying the problem of global warming and the scientific methods which are employed for the prediction of climate change due to the increases in greenhouse gases. The basic physics of the greenhouse effect is well understood. If atmospheric carbon dioxide concentration doubles and nothing else changes apart from atmospheric temperature, then the average global temperature near the surface will increase by about 1.2 °C. That figure is not disputed among scientists.

However, the situation is complicated by feedbacks and regional variations. Numerical models run on computers are the best tools available for addressing these problems. Although highly complex and at a relatively early stage of development, climate models are already capable of giving useful information of a predictive kind. Confidence in the models comes from the considerable skill with which they reproduce present climate and its variations (including perturbations such as the Pinatubo volcanic eruption) and also from the success of the few attempts which have been made to reproduce past climates; these latter are limited as much by the lack of data as by the inadequacies of the models.

However, model limitations remain, which give rise to uncertainty (see box). The predictions presented in Chapter 6 reflected these uncertainties, the largest of which are due to the models' failure to deal adequately with clouds and with the effects of the ocean circulation. Factors – such as the regional patterns of changes in rainfall – that most influence the impact of climate change are as yet the most uncertain.

With uncertainty in the basic science of climate change and in the predictions of future climate, there are bound also to be uncertainties in our assessment of the impact of climate change. As Chapter 7 shows, however, some important general statements can be made with reasonable confidence. Under a

---

The reasons for scientific uncertainty

The Intergovernmental Panel on Climate Change[1] has described the scientific uncertainty as follows.

'There are many uncertainties in our predictions particularly with regard to the timing, magnitude and regional patterns of climate change, due to our incomplete understanding of:

- sources and sinks of greenhouse gases, which affect predictions of future concentrations,
- clouds, which strongly influence the magnitude of climate change,
- oceans, which influence the timing and patterns of climate change,
- polar ice-sheets which affect predictions of sea level rise.

These processes are already partially understood, and we are confident that the uncertainties can be reduced by further research. However, the complexity of the system means that we cannot rule out surprises.'

---

business-as-usual scenario of increasing carbon dioxide emissions next century, the rate of climate change is likely to be large, probably greater than the Earth has seen for many millennia. Many ecosystems (including human beings) may not be able to adapt easily to such a rate of change. The most noticeable impacts are likely to be on the availability of water (especially on the frequency and severity of droughts and floods), on the distribution (though possibly not on the overall size) of global food production and on sea level in low-lying areas of the world. Further, although most of our predictions have been limited in range to the end of next century, it is clear that by the century beyond 2100 the magnitude of the change in climate and the impacts resulting from that change are likely to be very large indeed.

Rather less confidence is placed in estimates of the likely climate change in various broadly defined regions of the world. These estimates have been coupled with studies of the sensitivity to different climates of these regions' resources, such as water and food, and have enabled some assessment of impact to be carried out. 'Local' detail (still on a larger scale than the size of many small countries) has, however, yet to be filled in. Predictions of the detailed impact on resources for more local regions await in their turn better scientific predictions of the likely regional climate change. The absence of more certainty about local change makes it particularly hard for politicians and decision makers to know what is the appropriate and responsible action to take.

The IPCC Assessment

Because of the scientific uncertainty, it has been necessary to make a large effort to obtain the best assessment of present knowledge and to express it as clearly as possible. For these reasons the Intergovernmental Panel on Climate Change (IPCC) was set up jointly by two United Nations' bodies, the World

Meteorological Organization (WMO) and the United Nations Environmental Programme (UNEP). The IPCC's first meeting in November 1988 was very timely; it was held just as strong political interest in global climate change was beginning to develop. The Panel realized the urgency of the problem and established three working groups, one to deal with the science of climate change, one with impacts and a third one to deal with policy responses.

The task of the Science Assessment Working Group, of which I have been the chairman (since 1992, the co-chairman), has been to present in the clearest possible terms our knowledge of the science of climate change together with our best estimate of the climate change next century which is likely to occur as a result of human activities. The Working Group has produced four reports[2], a comprehensive report covering the whole field in 1990; two supplementary reports respectively in 1992 and 1994 addressing particular issues and a second comprehensive assessment in 1995. Previous chapters have already referred widely to these reports. I would like here to say more about how they were produced.

In preparing these reports we realized from the start that if they were to be really authoritative and taken seriously, it would be necessary to involve as many as possible of the world scientific community in their production. A small international organizing team was set up at the Hadley Centre of the United Kingdom Meteorological Office at Bracknell and through meetings, workshops and a great deal of correspondence most of those scientists in the world (both in universities and government-supported laboratories) who are deeply engaged in research into the science of climate change were involved in the preparation and writing of the reports. For the first report, 170 scientists from 25 countries contributed and a further 200 scientists were involved in its peer review. For the second comprehensive report in 1995, over 400 scientists from 26 countries submitted draft text and over 500 reviewers from 40 countries participated in its peer review.

In addition to the comprehensive, thorough and intensively reviewed background chapters which form the basic material for each assessment, each report includes a Summary for Policymakers (SPM), the wording of which is approved in detail at a plenary meeting of the Working Group, the object being to reach agreement on the science and on the best way of presenting the science to policymakers with accuracy and clarity. The plenary meeting which agreed unanimously the 1995 SPM was attended by 177 delegates from 96 countries, representatives from 14 non-governmental organizations and 28 lead authors of the scientific chapters. There has been very lively discussion at these plenary meetings, most of which has concerned achieving the most informative and accurate wording rather than because of dispute over scientific content. I remember, at the meeting at Guangzhou in China which agreed the text of the 1992 report, spending over three hours on the wording of one sentence, the meeting being very concerned to present that piece of information, including the degree of scientific uncertainty we felt regarding it, as clearly and as unambiguously as possible.

During the preparation of the reports, a considerable part of the debate amongst the scientists has centred on just how much can be said about the likely climate change next century. Particularly to begin with, some felt that the uncertainties were such that scientists should refrain from making any esti-

159

mates or predictions for the future. However, it soon became clear that the responsibility of scientists to convey the best possible information could not be discharged without making estimates of the most likely magnitude of the change next century coupled with clear statements of our assumptions and the level of uncertainty in the estimates. Weather forecasters have a similar, although much more short-term responsibility. Even though they may feel uncertain about tomorrow's weather, they cannot refuse to make a forecast. If they do refuse, they withhold from the public most of the useful information they possess. Despite the uncertainty in a weather forecast it provides useful guidance to a wide range of people. In a similar way the climate models, although subject to uncertainty, provide useful guidance for policy.

I have given these details of the work of the Science Assessment Group in order to demonstrate the degree of commitment of the scientific community to the understanding of global climate change and to the communication of the best scientific information to the world's politicians and policymakers. After all, the problem of global environmental change is one of the largest problems facing the world scientific community. No previous scientific assessments on this or any other subject have involved so many scientists so widely distributed both as regards their countries and their scientific disciplines. The IPCC reports can therefore be considered as authoritative statements of the contemporary views of the international scientific community.

A further important strength of the IPCC is that, because it is an intergovernmental body, governments are involved in its work. In particular, government representatives assist in making sure that the presentation of the science is both clear and relevant from the point of view of the policymaker. Having been part of the process, the resulting assessments are in a real sense owned by governments as well as by scientists – an important factor when it comes to policy negotiations.

In the presentation of the IPCC Assessments to politicians and policymakers, the degree of scientific consensus which has been achieved has been of great importance in persuading them to take seriously the problem of global warming and its impact. In the run-up to the United Nations Conference on Environment and Development (UNCED) at Rio de Janeiro in June 1992, the fact that they accepted the reality of the problem led to the formulation of the Climate Convention. It has often been commented that without the clear message which came from the world's scientists, orchestrated by the IPCC, the world's leaders would never have agreed to sign the Climate Convention.

Since the publication of the reports the debate concerning the scientific findings has continued in the world's press. Many have commented favourably on their clarity and accuracy. A few scientists have criticized because they feel the reports have insufficiently emphasized the uncertainties; others have expressed their disappointment that they have not spelt out the potential dangers to the world more forcefully. The scientific debate continues as indeed it must; argument and debate are intrinsic to the scientific process.

I have illustrated the work of the IPCC by describing in some detail the activity of the Science Assessment Working Group. The other Working Groups of the IPCC have followed similar procedures and have dealt with the Impacts of Climate Change, with Adaptation and Mitigations strategies and with the

Economics and Social Dimensions of Climate Change. Contributions to their work have not only come from natural scientists; increasingly social scientists especially economists have become involved. In these social science areas much fresh ground has been broken as consideration has been given to questions of what, in the global context, might form the basis of appropriate political and economic response to climate change. The rest of this chapter and the following chapters will draw heavily on their work.

Narrowing the uncertainty

A key question for policymakers is, 'How long will it be before the scientists are more certain about the projections of likely climate change, in particular concerning the regional and local detail?' Because of the enormous complexity of the climate system, we cannot expect the progress to be extremely rapid. Better models are needed, which in their turn will require bigger and faster computers. Above all, much better observations of all aspects of the climate system are required to describe climate variations as they occur and to calibrate and validate climate models.

In the atmosphere, clouds and all aspects of the hydrological (water) cycle need to be better observed. And it is the major oceans of the world, which cover a large fraction of the Earth's surface and which are particularly poorly monitored at the present, where observations with much higher accuracy and more complete coverage are urgently needed. To this end, new methods of observing the ocean surface from space vehicles have recently been developed and new means of observing the interior of the ocean are urgently being pursued. But not only are better physical measurements required. To be able to predict the detailed increases of greenhouse gases in the atmosphere, the problems of the carbon cycle must be unravelled; for this much more comprehensive measurements of the biosphere in the ocean as well as that on land are needed.

Stimulated by internationally organized observing programmes such as the Global Climate Observing System (GCOS), space agencies around the world have plans in place for deploying Earth-orbiting platforms around the turn of the century which will make many new observations relevant to the problems of climate change (see box).

The vast increase seen recently in the public and political interest in the problem of climate change has stimulated a large increase in scientific activity. Through this we can expect a steady progress in our understanding. However, because the new observations mentioned above will take some years to realize and to analyse it will probably be almost a decade before large strides towards more complete certainty can be made and before the required detail on the regional and local scales can be provided.

Alongside the increased understanding and more accurate predictions of likely climate change coming from the community of natural scientists, much more effort is also going into studies of different human activities and how they might be affected. Much better quantification of the impacts of climate change will result from these studies. Economists and other social scientists are also

Space observations of the climate system

For forecasting the weather round the world – for airlines, for shipping, for many other applications and for the public – meteorologists rely extensively on observations from satellites. Under international agreements, five geostationary satellites are spaced around the equator for weather observation; moving pictures from them have become familiar to us on our television screens. Information from polar orbiting satellites flown by the United States is also available to the weather services of the world to provide input into computer models of the weather and to assist in forecasting (see for instance Fig. 5.4).

These weather observations provide a basic input to climate models. But for climate prediction and research, observations from other components of the climate system, in particular from the oceans, are required. The ERS-1 satellite launched by the European Space Agency in 1991 is an example of a new generation of large satellites in which the latest techniques are directed to observing the Earth. It carries a payload particularly aimed at ocean observation (Fig. 9.1) including instruments for accurate measurement of sea surface temperature, the surface wind over the oceans (by means of a radar scatterometer) and the topography of the ocean surface (by means of a radar altimeter). This latter instrument can detect changes in the mean height of areas of the ocean surface with a precision of a few centimetres, enabling ocean currents to be located and measured. In addition ERS-1 carries a synthetic aperture radar (SAR) which provides images of the Earth's surface, including for instance its ice cover, and which is able to penetrate the cover of underlying clouds.

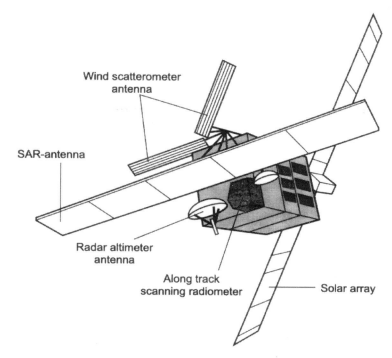

Wind scatterometer antenna

SAR-antenna

Radar altimeter antenna

Along track scanning radiometer

Solar array

Fig. 9.1
The ERS-1 satellite showing the solar array, the SAR antenna, the radar altimeter and scatterometer antennae and the along track scanning radiometer for accurate sea surface temperature measurement. The SAR antenna is 10 metres by 1 metre and the total mass of the satellite is 2.4 tonnes.

beginning to carry out detailed work on possible response strategies and the economic and political measures which will be necessary to achieve them. A lot of progress can be expected during the next decade.

Sustainable development

So much for uncertainty in the science of global warming. But how does this uncertainty map on to the world of political decision making?

One of the remarkable movements of the last few years is the way in which problems of the global environment have moved up the political agenda. In her speech at the opening in 1990 of the Hadley Centre at the United Kingdom Meteorological Office, Margaret Thatcher, the former British Prime Minister, explained our clear responsibility to the environment. 'We have a full repairing lease on the Earth. With the work of the IPCC, we can now say we have the surveyor's report; and it shows there are faults and that the repair work needs to start without delay. The problems do not lie in the future, they are here and now: and it is our children and grandchildren, who are already growing up, who will be affected.' Many other politicians have similarly expressed their feelings of responsibility for the global environment. Without this deeply felt and widely held concern, the UNCED conference at Rio, with environment as the number one item on its agenda, could never have taken place.

But, despite its importance, even when concentrating on the long term, the environment is only one of many considerations politicians must take into account. For developed countries, the maintenance of living standards, full employment (or something close to it) and economic growth have become dominant issues. Many developing countries are facing acute problems in the short term: basic survival and large debt repayment; others, under the pressure of large increases in population, are looking for rapid industrial development. However, an important characteristic of environmental problems, compared with many of the other issues faced by politicians, is that they are long-term and potentially irreversible – which is why Tim Wirth, the Under Secretary of State for Global Affairs in the United States Government has said, 'The economy is a wholly-owned subsidiary of the environment'.

A balance, therefore, has to be struck between the provision of necessary resources for development and the long term need to preserve the environment. That is why the Rio conference was about Environment and Development. The formula which links the two is called sustainable development (see box) – development which does not carry with it the overuse of irreplaceable resources or irreversible environmental degradation.

The idea of sustainable development echoes what was said in Chapter 8, when addressing more generally the relationship of humans to their environment and especially the need for balance and harmony. The Climate Convention signed at the Rio Conference also recognized the need for this balance. In the statement of its objective (see box in Chapter 10), it states the need for stabilization of greenhouse gas concentrations in the atmosphere. It goes on to explain that this should be at a level and on a timescale such that ecosystems are allowed to adapt to climate change naturally, that food production is not

---

Sustainable development

A number of definitions of sustainable development have been produced. The following two well capture the idea.

According to the Bruntland Commission Report *Our Common Future* presented in 1987, sustainable development is 'meeting the needs of the present without compromising the ability of future generations to meet their own needs'.

A more detailed definition is contained in the White Paper *This Common Inheritance*, published by the United Kingdom Department of the Environment in 1990: 'sustainable development means living on the Earth's income rather than eroding its capital' and 'keeping the consumption of renewable natural resources within the limits of their replenishment'. It recognizes the intrinsic value of the natural world explaining that sustainable development 'means handing down to successive generations not only man-made wealth (such as buildings, roads and railways) but also natural wealth, such as clean and adequate water supplies, good arable land, a wealth of wildlife and ample forests'.

The United Kingdom Government's first strategy report on sustainable development, issued in January 1994[3], defined four principles which should govern necessary collective action:

- Decisions should be based on the best possible scientific information and analysis of risks.
- Where there is uncertainty and potentially serious risks exist, precautionary action may be necessary.
- Ecological impacts must be considered, particularly where resources are non-renewable or effects may be irreversible.
- Cost implications should be brought home directly to the people responsible – the 'polluter pays' principle.

---

threatened and that economic development can proceed in a sustainable manner.

Why not wait and see?

In the light of the scientific uncertainty, it is often argued that the case is not strong enough for any action to be taken now. What we should do is to obtain as quickly as possible, through appropriate research programmes, much more precise information about future climate change and its impact. We would then, so the argument goes, be in a much better position to decide on relevant action.

It is true that more accurate information is urgently needed so that decisions can be better informed. But in any sensible future planning, all information about the likely future needs to be taken properly into account. Decisions now should be informed by the best information available now, even if it is imperfect.

In the first place, quite a lot is already known – enough to scope the problem as a whole. There is general consensus amongst scientists about the most likely overall magnitude of climate change and there are good indications about its probable impact. Although we are not yet very confident regarding detailed predictions, enough is known to realize that the rate of climate change due to increasing greenhouse gases will almost certainly bring substantial deleterious effects and pose a large problem to the world. It will hit some countries much more than others. Those worst hit are likely to be those in the developing world that are least able to cope with it. Some countries may actually experience a more beneficial climate. But in a world where there is increasing interdependence between nations, no nation will be immune from the effects.

Secondly, the timescales of both atmospheric and human responses are long. Carbon dioxide emitted into the atmosphere today will contribute to the increased concentration of this gas and the associated climate change for over a hundred years. The more that is emitted now, the more difficult it will be to reduce atmospheric carbon dioxide concentration to the levels which will eventually be required. With regard to human response, the major changes which are likely to be required, for instance in large-scale infrastructure, will take many decades. Large power stations which will produce electricity in thirty or forty years' time are being planned and built today. The demands which are likely to be placed all of us because of concerns about global warming need to be brought into the planning process now.

Thirdly, many of the required actions not only lead to substantial reductions in greenhouse gas emissions but they are good to do for other reasons which bring other direct benefits – such proposals for action are often described as 'no regrets' proposals.

Many actions addressing increased efficiency lead also to net savings in cost (sometimes called 'win-win' measures). Other actions lead to improvements in performance or additional comfort.

Fourthly, there are more general beneficial reasons for some of the proposed actions. In Chapter 8 it was pointed out that humans are far too profligate in their use of the world's resources. Fossil fuels are burnt and minerals are used, forests are cut down and soil is allowed to be eroded, without any serious thought of the needs of future generations. The imperative of the global warming problem will help us to use the world's resources in a more sustainable way. Further, the technical innovation which will be required in the energy industry – in energy efficiency and conservation and in renewable energy development – will provide a challenge and opportunity to the world's industry to develop important new technologies – more of that in Chapter 11.

The Precautionary Principle

Some of these arguments for action are applications of what is often called the Precautionary Principle, one of the basic principles which was included in the Rio Declaration at the Earth Summit in June 1992. Principle 15 in the Declaration reads, 'In order to protect the environment, the precautionary approach shall be widely applied by States according to their capabilities. Where there are threats of serious or irreversible damage, lack of full scientific certainty shall not be used as a reason for postponing cost-effective measures to prevent environmental degradation.' A similar statement is contained in article 3 of the Climate Convention (see box in Chapter 10).

We often apply the Precautionary Principle in our day-to-day living. We take out insurance policies to cover the possibility of accidents or losses; we carry out precautionary maintenance on housing or on vehicles, and we readily accept that in medicine prevention is better than cure. In all these actions we weigh up the cost of insurance or other precautions against the possible damage and conclude that the investment is worthwhile. The arguments are very similar as the Precautionary Principle is applied to the problem of global warming.

In taking out an insurance policy we often have in mind the possibility of the really unexpected. Although covering ourselves for the most unlikely happenings is not our main reason for taking out the insurance, our peace of mind is considerably increased if the policy includes these improbable events. In a similar way, in arguing for action concerning global warming, some have strongly emphasized the need to guard against the possibility of surprises. They point out that, because of positive feedbacks which are not yet well understood[4], the increase of some greenhouse gases could be much larger than is currently predicted. They also point to the evidence that rapid changes of climate have occurred in the past (Fig. 4.6 and 4.7) possibly because of dramatic changes in ocean circulation; they could presumably occur again.

The risk posed by such possibilities is impossible to assess. It is, however, salutary to call attention to the discovery of the ozone 'hole' over Antarctica in 1985. Scientific experts in the chemistry of the ozone layer were completely taken by surprise by that discovery. In the years since its discovery, the 'hole' has substantially increased in depth. Resulting from this knowledge, international action to ban ozone-depleting chemicals has progressed much more rapidly. The lesson for us here is that the climate system may be more vulnerable to disturbance than we have often thought it to be. When it comes to future climate change, it would not be prudent to rule out the possibility of surprises.

When selling their policies, insurance companies often trade on our fear of the unexpected; when faced with substantial uncertainty it is easy to home in on the possibility of the really unknown, especially the more devastating possibilities. However, in weighing the action that needs to be taken with regard to future climate change, although the possibility of surprises should not be ignored, neither should they feature as the main argument for action. Much stronger in the argument for precautionary action is the realization that significant anthropogenic climate change is not an unlikely possibility but a near certainty; it is no change of climate that is unlikely. The uncertainties which mainly have to be weighed lie in the magnitude of the change and the details of its regional distribution.

An argument which is sometimes advanced for doing nothing now is that by the time action is really necessary, more technical options will be available. By acting now, we might foreclose their use. Any action taken now must, of course, take into account the possibility of helpful technical developments. But the argument also works the other way. The thinking and the activity generated by considering appropriate actions now and by planning for more action later will itself be likely to stimulate the sort of technical innovation which will be required.

While speaking of technical options, I should briefly mention possible options to counteract global warming by the artificial modification of the environment (sometimes referred to as geoengineering[5]). A number of proposals for 'technical fixes' of this kind have been put forward, for instance: the installation of mirrors in space to cool the Earth by reflecting sunlight away from it; the addition of dust to the upper atmosphere to provide a similar cooling effect and the alteration of cloud amount and type by adding cloud condensation nuclei to the atmosphere. None of these have been demonstrated either to be feasible or effective. Further, they suffer from the very serious problem that none of them would exactly counterbalance the effect of increasing greenhouse gases. As has been shown, the climate system is far from simple. The results of any attempt at large-scale climate modification could not be perfectly predicted and might not be what is desired. With the present state of knowledge, artificial climate modification along any of these lines is not an option that need be considered.

The conclusion from this section – and the last one – is that to 'wait and see' would be an inadequate and irresponsible response to what we know. The Climate Convention signed in Rio (see box in Chapter 10) recognized that some action needs to be taken now. Just what that action should be and how it fits in to a sensible scheme of sequential decision making will be the subject of the next chapter.

Some global economics

So far in this chapter, our attempt to balance uncertainty against the need for action has been considered in terms of issues. Is it possible to carry out the weighing in terms of cost? In a world which tends to be dominated by economic arguments, quantification of the costs of action against the likely costs of the consequences of inaction must at least be attempted. It is also helpful to put these costs in context by comparing them with other items of global expenditure.

At the end of Chapter 7 estimates of the cost of the damage from global warming were presented. These were based on the assumption that, resulting from human activities, the increase in greenhouse gases in the atmosphere would be equivalent to a doubling of the carbon dioxide concentration – a situation which, under business-as-usual would occur before the middle of next century. The estimates were typically around 1–1.5 per cent of GDP for developed countries. In developing countries, because of their greater vulnerability to climate change and because a greater proportion of their expenditure is dependent on activities such as agriculture and water, estimates of the cost of the damage are greater, typically about 5 per cent of GDP (range from 2 to 9 per cent). Averaged over world as a whole, estimates typically fall in the range 1.5 to 2 per cent of globally aggregated GDP (sometimes called Gross World Product or GWP). Although, at the present stage of knowledge, these estimates are bound to be crude, nevertheless they give a feel for the likely range of cost.

The longer-term damage, should greenhouse gases more than double in concentration, is likely to rise somewhat more steeply in relation to the concentration of carbon dioxide (Fig. 9.2). For quadrupled equivalent carbon dioxide concentration, for instance, estimates of damage cost of the order of 4–6 per cent of GWP or more have been made – suggesting that the damage might follow something like a quadratic law relative to the expected temperature rise[6]. In addition the much larger degree of climate change would considerably enhance the possibility of surprises and irreversible change.

Since the main contribution to global warming arises from carbon dioxide emissions, attempts have also been made to express these costs in terms of the cost per tonne of carbon as carbon dioxide emitted from human activities. A simple, but crude calculation can be carried out as follows. Consider the situation when the carbon dioxide concentration in the atmosphere has doubled from its pre-industrial value, which will occur when an additional amount of carbon as carbon dioxide of about 800 Gt from anthropogenic sources has been emitted into the atmosphere (cf Fig. 3.1 and recall that about half the carbon dioxide emitted accumulates in the atmosphere). This carbon dioxide will remain in the atmosphere on average for of the order of 100 years. Assuming a figure of 2 per cent of GWP – or 400 billion $(US) per annum – as the cost of the damage due to global warming in that situation, and assuming also that the damage remains over the 100 years of the life-time of the carbon dioxide in the atmosphere, the cost per tonne of carbon turns out to be about 50$.

Calculations of the cost per tonne of carbon can be made with much more sophistication by taking into account that it is the *marginal* damage cost (that is the cost of the damage due to one extra tonne of carbon emitted now) that is really required and also by allowing through a discount rate for the fact that it is damage some time in the future that is being costed now. Estimates made by different economists then range from 5–125 US$[7] – the very large range being due the different assumptions which have been made. The numbers are particularly sensitive to the discount rate which is assumed; values at the top end of the range above about 50$ have assumed a discount rate of less than 2 per cent; those at the bottom end have assumed a discount rate around 5 per cent.

Fig 9.2
The shape of the curves of the damage costs of climate change and the mitigation costs as a function of atmospheric carbon dioxide emissions reductions[8]. The arrow shows the optimal reduction level.

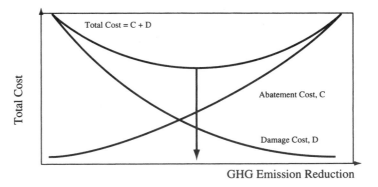

Amongst economists there is much debate but no agreement about how to apply discount accounting to this sort of problem or about what rate is most appropriate. However, for such long-term considerations a smaller discount rate seems more appropriate than a large one[9] (applying a 5 per cent discount rate, for instance, devalues costs 50 years into the future by a factor of 13). For our broad economic arguments in later chapters we shall therefore quote an estimate of damage cost in the range 50$ to 100$ per tonne of carbon emitted as carbon dioxide.

The alternative to adapting to global warming and bearing the cost is to avoid or mitigate its effects by reducing greenhouse gas emissions, in particular the emissions of carbon dioxide. The cost of mitigation is very dependent on the amount of reduction required in greenhouse gas emissions; large reductions will cost proportionately more than small ones (Fig 9.2). It will also depend on the timescale of reduction. To reduce emissions drastically in the very near term would inevitably mean large reductions in energy availability with significant disruption to industry and large cost. However, more gradual reductions can be made with relatively small cost through actions of two kinds. Firstly, substantial efficiency gains in the use of energy can easily be achieved many of which would lead to cost savings; these can be put into train now. Secondly, in the generation of energy, again proven technology exists for substantial efficiency improvements and also for the bringing into use renewable sources of energy generation which are not dependent on fossil fuels. These can be planned for now and changes made as energy infrastructure, which has a typical life of 30 years or so, becomes ready for replacement. The next two chapters will present more detail about these possible actions.

Our purpose here is to look at the likely overall cost of mitigation, much of which will arise in the energy or the transport sectors as cheap fossil fuels are replaced by other energy sources which, at least in the short term, are likely to be more expensive. A number of studies have been made which have attempted to take into account all the relevant factors although they are bound to be surrounded by substantial uncertainty. One of the most difficult factors to take into account is that of likely future innovation.It is not easy to peer into the crystal ball of technical development; almost any attempt to do so is likely to underestimate its future potential. For this reason the estimates of mitigation cost are probably on the high side.

In the next chapter we shall look in more detail at the reductions in emissions which are required to stabilize carbon dioxide concentrations in the atmosphere at different levels over the next one or two hundred years. A number of economic models have been applied to providing estimates of the cost to the world's economy of these reductions. They show, firstly that the timescale of the reductions has a large influence on the cost; attempting large reductions too soon is much more costly than allowing replacement technologies and plant to be installed more on their natural timescale. In Fig. 9.3 are shown three different profiles of emissions leading to atmospheric carbon dioxide concentrations of about 500 ppm at 2100 and for all of which the cumulative emissions up to the year 2100 are about the same. Calculations from two different economic models show that the cost of achieving profile *c* (emissions stabilization) is the largest at somewhat over 1 per cent of GWP and

Fig 9.3
Possible profiles of future carbon dioxide emissions from the combustion of fossil fuels[13]. The full curve is 'business as usual'; curves a and b allow for some increase in emissions during the first half of next century and show substantial reductions during the second half; curve c shows emissions stabilization.

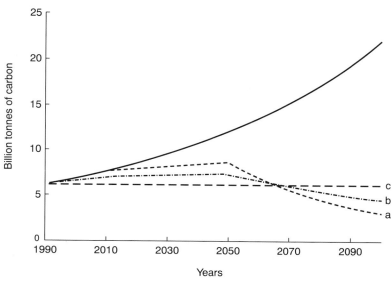

Fig 9.4
Estimates, with two different economic models, of mitigation costs as a function of different levels of stabilization of the atmospheric carbon dioxide concentration[11].

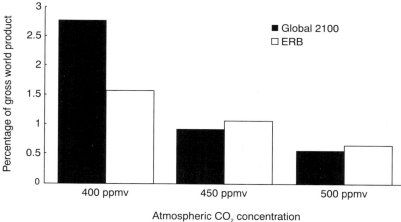

that achieving profile a (which allows for some increase in emissions now compensated by greater reductions later on) is the least costly at perhaps 0.6 per cent of GWP.

Secondly studies with the same models show (Fig. 9.4) that the cost is substantially dependent on the target level of stabilization of atmospheric carbon dioxide concentration. However, for all stabilization levels above 450 ppmv, the cost is estimated to be below 1 per cent of GWP. More elaborate models called Integrated Assessment Models (IAMs) – see box – are currently being built to address problems of this kind in a more complete manner.

However, it should be noted that, even if the carbon dioxide concentration is stabilized at 400 or 450 ppmv, remembering that the effect of increases in the other greenhouse gases also has to be included (see chapter 10, page 184), the world will have been committed to a significant degree of climate change,

Integrated Assessment Models[12]

Integrated Assessment Models or IAMs are models which represent within one integrated numerical model (Fig. 9.5) the physical, chemical and biological processes which control the concentration of greenhouse gases in the atmosphere, the physical processes which determine the effect of changing greenhouse gas concentrations on climate and sea level, the biology and ecology of ecosystems (natural and managed), the physical and human impacts of climate change and the socio-economics of adaptation to and mitigation of climate change. Such models are highly sophisticated and complex although their components are bound to be very simplified.

The main purpose of such models is to study the connections and the interactions between the various components. Because of their complexity and because of the non-linear nature of many of the interactions, a great deal of care and skill is needed in the interpretation of the results from such models. In 1995, upwards of 25 such models were being run in different laboratories and institutes around the world.

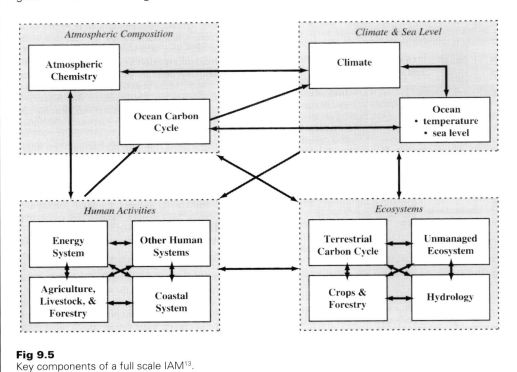

Fig 9.5
Key components of a full scale IAM[13].

bringing with it substantial costs and demands for adaptation. What is being mitigated is further climate change.

Such 'weighing' of the economics as has been possible so far therefore brings four messages. The first is that early action at small cost can be undertaken now to reduce emissions and to slow the rate of change. The second is that, in the longer term, as presently estimated, the cost in economic terms of mitigation and avoidance is less in global average terms than the likely cost of the damage from global warming (including only that part of the damage for which estimates of cost in money terms can be provided) or of adapting to it. The third message is that drastic action now to stabilize carbon dioxide concentration at close to today's level within the next few decades – and therefore largely to eliminate climate change now – would be expensive compared to the cost of action to stabilize the carbon dioxide concentration on a somewhat longer time-scale (say by the end of next century) and at a somewhat higher level. The fourth message is that, by the time adequate action is taken, significant climate change will already have occurred bringing with it substantial demands for adaptation. Since the weighing of economic factors is an important consideration for politicians and policymakers, work to refine these estimates of climate change, damage and economic cost needs to be pursued with urgency. We shall return to this issue in the context of the international Framework Convention on Climate Change towards the end of the next chapter.

In considering both the cost of the impact of global warming and the cost of adaptation or of mitigation, figures of the order of a few per cent of GDP have been mentioned. It is interesting to compare this with other items of expenditure in national or personal budgets. In a typical developed country, for example the United Kingdom, about 5 per cent of personal income is spent directly on the supply of primary energy (basic fuel such as coal, oil and gas, fuel for electricity supply and fuel for transport), about 6 per cent on health and 4 to 5 per cent on defence. It is, of course, clear that global warming is strongly linked to energy production – it is because of the way energy is provided that the problem exists – and this subject will be expanded in the next two chapters. But the impacts of global warming also have implications for health – such as the possible spread of disease – and for national security – for example, the possibility of wars fought over water, or the impact of large numbers of environmental refugees. Any thorough consideration of the economics of global warming needs therefore to assess the strength of these implications and to take them into account in the overall economic balance.

So far, on the global warming balance sheet we have estimates of costs and of benefits or drawbacks. What we do not have as yet is a capital account. Valuing human-made capital is commonplace, but in the overall accounting we are attempting, 'natural' capital must clearly be valued too. By 'natural' capital I mean, for instance, natural resources which may be renewable (such as a forest) or non-renewable (such as coal, oil or minerals)[14]. Their value is clearly more than the cost of exploitation or extraction.

Other items, some of which were mentioned at the end of Chapter 7, such as natural amenity and the value of species, can also be considered as 'natural' capital. I have argued (Chapter 8) that there is intrinsic value in the natural world – indeed, the value and importance of such 'natural' capital is increasingly recognized. The difficulty is that it is neither possible nor appropriate to express much of this value in money. Despite this difficulty, it is now widely recognized that national and global indicators of sustainable development should be prepared which include items of 'natural' capital and ways of including such items in national balance sheets are being actively pursued.

In summary, therefore, the items in the overall global warming balance sheet are:

- estimates (with considerable uncertainty) of cost (for those items which can be quantified in terms of money) of the likely impacts of global warming supposing the equivalent atmospheric carbon dioxide concentration were to double, which are typically about 1.5 to 2 per cent of GWP averaged over the world as a whole – 1 to 1.5 per cent of GDP in developed countries and typically 5 per cent (2 to 9 per cent) in developing countries.
- estimates of the cost of mitigation and avoidance of global warming – typically about 1 per cent of GWP or less, although the likely cost of drastic and immediate action to prevent any further climate change is likely to be larger.
- estimates of the cost of adaptation to the global warming which will inevitably occur before actions taken in mitigation are adequate to stabilize atmospheric greenhouse gas concentrations and prevent further anthropogenic climate change.
- estimates of the impacts of global warming which are difficult if not impossible to value in money terms, for instance, those with social consequences, those which affect human amenity and 'natural' capital or those which have implications for national security.

There is already international acceptance that some action to mitigate global warming is necessary. The next chapter will consider some of the actions in more detail but still in the wider context of the requirement for sustainable development.

Questions

1   It is sometimes argued that, in scientific enquiry, 'consensus' can never be achieved, because debate and controversy are fundamental to the search for scientific truth. Discuss what is meant by 'consensus' and whether you agree with this argument. Do you think the IPCC reports have achieved 'consensus'?

2   How much do you think the value of IPCC reports depends on (1) the peer review process to which they have been subjected, (2) the involvement of governments in the presentation of scientific results?

3   Look out as many definitions of 'Sustainable Development' as you can find. Discuss which you think is the best.

4   Make a list of appropriate indicators which might be used to assess the degree to which a country is achieving sustainable development. Which do you think might be the most valuable?

5   Work out the value of a 'cost' today if it is 20, 50 or 100 years into the future and the assumed discount rate is 1%, 2% or 5%. Look up and summarize the arguments for discounting future costs as presented in various chapters of the IPCC 1995 Report[15]. What do you think is the most appropriate discount rate to use?

6   Construct, as far as you are able, a set of environmental accounts for your country including items of 'natural' capital. Your accounts will not necessarily be all in terms of money.

Notes

1 *Climate Change, The IPCC Scientific Assessment*, eds. J. T. Houghton, G. J. Jenkins and J. J. Ephraums, CUP, 1990, p. 365; Executive Summary p. xii. Similar but more elaborate statements are in the 1995 Report.

2 *Climate Change, The IPCC Scientific Assessment*, eds. J. T. Houghton, G. J. Jenkins and J. J. Ephraums, CUP, 1990; *Climate Change 1992, The Supplementary Report to the IPCC Scientific Assessment*, eds. J. T. Houghton, B. A. Callendar and S. K. Varney, CUP, 1992; *Climate Change 1994*, eds. J. T. Houghton, L. G. Meira Filho, B. A. Callander, E. Haites, N. Harris and K. Maskell, CUP, 1995; *Climate Change 1995: The Science of Climate Change*, eds. J. T. Houghton, L. G. Meira Filho, B. A. Callender, N. Harris, A. Kattenberg and K. Maskell, CUP, 1996; *Climate Change 1995: Impacts, Adaptations and Mitigation of Climate Change*, eds. R. T. Watson, M. C. Zinyowera and R. H. Moss, CUP, 1996; and *Climate Change 1995: Economic and Social Dimensions of Climate Change*, eds. J. Bruce, Hoesung Lee and E. Haites, CUP, 1996.

3 *Sustainable Development: the UK Strategy*, London, HMSO, Cm 2426, 1994, p. 7.

4 See Chapter 3.

5 Reviewed in *Policy Implications of Greenhouse Warming*, National Academy Press, Washington DC, 1992, pp. 433–64.

6 D.W. Pearce *et al.*, chapter 6 in *Climate Change 1995: Economic and Social Dimensions of Climate Change*, eds. J. Bruce, Hoesung Lee and E. Haites, CUP, 1996.

7 *Climate Change 1995: Economic and Social Dimensions of Climate Change*, eds. J. Bruce, Hoesung Lee and E. Haites, CUP, 1996, Summary for Policymakers.

8 From M. Munasinghe *et al.*, 'Applicability of techniques of cost–benefit analysis to climate change, chapter 5 in *Climate Change 1995: Economic and Social Dimensions of Climate Change*, eds. J. Bruce, Hoesung Lee and E. Haites, CUP, 1996.

9 See W. R. Cline *'The Economics of Global Warming'* Institute for International Economics, Washington DC, 1992, chapter 6.

10 From R. Richels and J. Edmonds, 'The economics of stabilising atmospheric CO_2 concentrations', *Energy Policy*, **23**, 1995, pp. 373–9.

11 From R. Richels and J. Edmonds 'The economics of stabilising atmospheric CO_2 concentrations', *Energy Policy*, **23**, 1995, pp. 373–9.

12 J. Weyant *et al.*, 'Integrated assessment of climate change', chapter 10 in *Climate Change 1995: Economic and Social Dimensions of Climate Change*, eds. J. Bruce, Hoesung Lee and E. Haites, CUP, 1996.

13 J. Weyant *et al.*, 'Integrated assessment of climate change', chapter 10 in *Climate Change 1995: Economic and Social Dimensions of Climate Change*, eds. J. Bruce, Hoesung Lee and E. Haites, CUP, 1996.

14 For a discussion of this issue see H. E. Daly, 'From empty-world economics to full-world economics: a historical turning point in economic development' in *World Forests for the Future*, eds. K. Ramakrishna and G. M. Woodwell, Yale University Press, 1993, pp. 79–91.

15 *Climate Change 1995: Economic and Social Dimensions of Climate Change*, eds. J. Bruce, Hoesung Lee and E. Haites, CUP, 1996.

10 A Strategy for Action to Slow and Stabilize Climate Change

Following the awareness of the problems of climate change aroused by the IPCC scientific assessments, the necessity of international action has been recognized. In this chapter I address the forms that action could take.

The Climate Convention

The United Nations Framework Convention on climate change signed by over 160 countries at the United Nations Conference on Environment and Development held in Rio de Janeiro in June 1992 came into force on 21 March 1994. It has set the agenda for action to slow and stabilize climate change. The signatories to the Convention (some of the detailed wording is presented in the box below) recognized the reality of global warming, recognized also the uncertainties associated with current predictions of climate change, agreed that action to mitigate the effects of climate change needs to be taken and pointed out that developed countries should take the lead in this action.

The Convention mentions one particular aim concerned with the relatively short term and one far reaching objective. The particular aim is that developed countries (Annex I countries in Climate Convention parlance) should take action to return greenhouse gas emissions, in particular those of carbon dioxide, to their 1990 levels by the year 2000. The long-term objective of the Convention, expressed in Article 2, is that the concentrations of greenhouse gases in the atmosphere should be stabilized 'at a level which would prevent dangerous anthropogenic interference with the climate system', the stabilization to be achieved within a time-frame sufficient to allow ecosystems to adapt naturally to climate change, to ensure that food production is not threatened and to enable economic development to proceed in a sustainable manner. In setting this objective, the Convention has recognized that it is only by stabilizing the concentration of greenhouse gases (especially carbon dioxide) in the atmosphere that the rapid climate change which is expected to occur with global warming can be halted.

Two sessions of the Conference of the Parties to the Climate Convention have so far taken place, in March/April 1995 and July 1996. A third is planned to be held in Kyoto, Japan towards the end of 1997 at which the main discussion will focus around commitments which can be made for greenhouse gas reductions during the early decades of next century.

The following paragraphs will first outline the short-term actions which are possible to begin to meet the requirements of the Convention. Further actions necessary to satisfy the Convention's objective to stabilize greenhouse gas concentrations will then be considered.

Some extracts from the United Nations Framework Convention on climate change, signed by over 160 countries in Rio de Janeiro in June 1992

First, some of the paragraphs in its preamble, where the parties to the Convention:

CONCERNED that human activities have been substantially increasing the atmospheric concentration of greenhouse gases, that these increases enhance the natural greenhouse effect, and that this will result on average in an additional warming of the Earth's surface and atmosphere and may adversely affect natural ecosystems and humankind,

NOTING that the largest share of historical and current global emissions of greenhouse gases has originated in developed countries, that per capita emissions in developing countries are still relatively low and that the share of global emissions originating in developing countries will grow to meet their social and development needs,

RECOGNIZING that various actions to address climate change can be justified economically in their own right and can also help in solving other environmental problems,

RECOGNIZING that low-lying and other small island countries, countries with low-lying coastal, arid and semi-arid areas or areas liable to floods, drought and desertification, and developing countries with fragile mountainous ecosystems are particularly vulnerable to the adverse effects of climate change,

AFFIRMING that responses to climate change should be coordinated with social and economic development in an integrated manner with a view to avoiding adverse impacts on the latter, taking into full account the legitimate priority needs of developing countries for the achievement of sustained economic growth and the eradication of poverty,

DETERMINED to protect the climate system for present and future generations, have AGREED as follows:

The Objective of the Convention is contained in Article 2 and reads as follows:

'The ultimate objective of this Convention and any related legal instruments that the Conference of the Parties may adopt is to achieve, in accordance with the relevant provisions of the Convention, stabilization of greenhouse gas concentrations in the atmosphere at a level that would prevent dangerous anthropogenic interference with the climate system. Such a level should be achieved within a time frame sufficient to allow ecosystems to adapt naturally to climate change, to ensure that food production is not threatened and to enable economic development to proceed in a sustainable manner.'

Article 3 deals with principles and includes agreement that the Parties 'take precautionary measures to anticipate, prevent or minimize the causes of climate change and mitigate its adverse effects. Where there are threats of serious or irreversible damage, lack of full scientific certainty should not be used as a reason for postponing such measures, taking into account that policies and measures to deal with climate change should be cost-effective so as to ensure global benefits at the lowest possible cost.'

Article 4 is concerned with Commitments. In this article, each of the signatories to the Convention agreed 'to adopt national policies and take corresponding measures on the mitigation of climate change, by limiting its anthropogenic emissions of greenhouse gases and protecting and enhancing its greenhouse sinks and reservoirs. These policies and measures will demonstrate that developed countries are taking the lead in modifying longer-term trends in anthropogenic emissions consistent with the objective of the Convention, recognizing that the return by the end of the present decade to earlier levels of anthropogenic emissions of carbon dioxide and other greenhouse gases not controlled by the Montreal Protocol would contribute to such modification ...'

Each signatory also agreed 'in order to promote progress to this end ... to communicate ... detailed information on its policies and measures referred to above, as well as on its resulting projected anthropogenic emissions by sources and removals by sinks of greenhouse gases not covered by the Montreal Protocol ... with the aim of returning individually or jointly to their 1990 levels these ... emissions ...'

Stabilization of emissions

The target for short-term action proposed for developed countries by the Climate Convention is that, by the year 2000, greenhouse gas emissions should be brought back to no more than their 1990 levels. In the run-up to the Rio conference, before the Climate Convention was formulated, many developed countries had already announced their intention to meet such a target at least for carbon dioxide. They would do this mainly through energy-saving measures, through switching to fuels such as natural gas, which for the same energy production generates 40 per cent less carbon dioxide than coal and 30 per cent less than oil. In addition those countries with traditional heavy industries (eg the iron and steel industry) are experiencing large changes which significantly reduce fossil fuel use. More detail of these energy-saving measures are given in the next chapter, which is devoted to a discussion of future energy needs and production.

By 1995, halfway through the decade, energy related emissions of carbon dioxide from those developed countries belonging to the OECD (Organisation of Economic Cooperation and Development) had increased in total by about 4 per cent from their 1990 values. For the rest of the world, carbon dioxide emissions from the Economies in Transition in the countries of the old Soviet Union had decreased by around 25 per cent because of the collapse of their economies and the total of emissions from developing countries had increased by around 15 per cent. Added together for the world as a whole during this period energy related carbon dioxide emissions remained almost constant.

For the second half of the decade, in some developed countries it is clear that emissions will continue to rise and that many will fail to reach the target of constraining emissions in 2000 to 1990 levels. Developing countries' emissions will continue to rise steadily and those from economies in transition cannot be expected to fall further.

Beyond the year 2000 a number of proposals have been made for emissions reductions by developed countries. The Alliance of Small Island States (AOSIS), a group particularly vulnerable to sea-level rise, has asked for a 20 per cent reduction from 1990 levels by 2005; the European Union has suggested a 15 per cent reduction by 2010, again relative to 1990, for its members taken together; Denmark has proposed a 50 per cent reduction by 2030; the Netherlands has proposed continuing reductions of one or two per cent per annum. The various possibilities will be addressed by the 1997 Conference of Parties of the Climate Convention. What is crucial is not only that reductions for the next decade or two are considered but also that such reductions are put in the context of the more substantial longer-term reductions that are likely to be necessary; we shall be discussing them later in the chapter.

The Montreal Protocol

The chlorofluorocarbons (CFCs) are greenhouse gases whose emissions into the atmosphere are already controlled under the Montreal Protocol on ozone depleting substances. This control has not arisen because of their potential as greenhouse gases, but because they deplete atmospheric ozone (see Chapter 3). Emissions of CFCs have fallen sharply during the last few years and the growth in their concentrations has slowed; for some CFCs a slight decline in their concentration is now apparent. The phase-out of their manufacture in industrialized countries by 1996 and in developing countries by 2006 as required by the 1992 amendments to the Montreal Protocol will ensure that the profile of their atmospheric concentration will continue to decline. However, because of their long life in the atmosphere this decline will be slow, it will be a century or more before their contribution to global warming is reduced to a negligible amount.

The replacements for CFCs, the HCFCs, which are also greenhouse gases – though less potent than the CFCs – are required to be phased out by 2030. It will probably be close to that date before their atmospheric concentration stops rising and begins to decline.

Because of the international agreements which now exist for control of the production of the CFCs and many of the related species which contribute to the greenhouse effect, for these gases the stabilization of atmospheric concentration required by the Climate Convention will in due course be achieved.

Other replacements for CFCs are the HFCs, which are greenhouse gases but not ozone-depleting. The controls of the Montreal Protocol do not therefore apply and, as was mentioned in Chapter 3, any substantial growth in the use of HFCs needs to be watched.

Forests

We now turn to the situation of the world's forests and the contribution that they can make to the mitigation of global warming. Action here can easily be taken now and is commendable for many other reasons.

Over the past few centuries many countries, especially those at mid-latitudes, have removed much of their forest cover to make room for agriculture. Many of the largest and most critical remaining forested areas are in the tropics. In recognition of the extremely valuable resource these areas represent, the developing countries where they lie are beginning to concentrate seriously on the management of their forests, on limiting the extent of deforestation or planning for substantial afforestation. Other large areas of forest lie at higher latitudes where developed countries can also take useful action to contribute to the alleviation of the problem of global warming.

During the last two decades, the additional needs of the increasing populations of developing countries for agricultural land and for fuelwood together with the rise in demand for tropical hardwoods by developed countries has led to a worrying rate of loss of forest in tropical regions. In many tropical countries the development of forest areas has been the only hope of subsistence for many people. Unfortunately, because the soils and other conditions were often inappropriate, much of this forest clearance has not led to sustainable agriculture but to serious land and soil degradation[1].

Measurements on the ground and observations from orbiting satellites have been combined to provide estimates of the area of tropical forest lost. Over the decade of the 1980s the average loss was about 1 per cent per year (Table 10.1),

Table 10.1
Estimates, from the United Nations Food and Agriculture Organization (FAO), of forest cover and deforestation for 87 countries in tropical regions[2]. The countries include almost all of the moist forest zone together with some dry areas. The figures are indicative and should not be taken as regional averages.

Continent	Forest Area 1980	Forest Area 1990	Annual deforestation 1981–90	Rate of change 1981–90 (% year)
	(thousands of square kilometres)			
Africa	6,500	6,000	50	−0.8
Latin America and Caribbean	9,230	8,400	83	−0.9
Asia	3,210	2,750	37	−1.2
Total	18,940	17,150	170	−0.9

although in some areas it was considerably higher. Such rates of loss cannot be sustained if much forest is to be left in fifty or a hundred years' time. The loss of forests is damaging, not only because of the ensuing land degradation but also because of the contribution that loss makes to global warming. There is also the dramatic loss in biodiversity (it is estimated that over half the world's species live in tropical forests) and the potential damage to regional climates (loss of forests can lead to a significant regional reduction in rainfall – see box on p. 121).

For every square kilometre of a typical tropical forest there are between about 20,000 and 50,000 tonnes of biomass (total living material), containing 10,000–25,000 tonnes of carbon[3]. It is estimated that burning or other destruction from deforestation turns about two-thirds of this carbon into carbon dioxide. On this basis, from the destruction of 170,000 square kilometres over the decade of the 1980s (Table 10.1) nearly 2 Gt (1 Gt = 1 gigatonne = 1,000 million tonnes) of carbon would enter the atmosphere as carbon dioxide. Although there are substantial uncertainties in the numbers, they tally with the IPCC estimate, quoted in Chapter 3 (see Table 3.1), of the carbon as carbon dioxide entering the atmosphere each year from tropical land-use change (mostly deforestation) of 1.6 ± 1.0 Gt per year – a significant fraction of the current total emissions of carbon dioxide into the atmosphere from human activities. Reducing deforestation can therefore make a substantial contribution to slowing the increase of greenhouse gases in the atmosphere, as well as the provision of other benefits such as guarding biodiversity and avoiding soil degradation.

So much for slowing deforestation. How about the possibilities for afforestation? For every square kilometre, a growing forest fixes between about 200 and 500 tonnes of carbon per year[4]. To illustrate the effect of afforestation on atmospheric carbon dioxide, suppose that an area of 100,000 square kilometres, a little more than the area of the island of Ireland, were planted each year for 40 years – starting now. By the year 2035, 4 million square kilometres would have been planted, that is roughly half the area of Australia. By the time the new forests matured – between 40 and 100 years after planting (the actual period depending on the type of forest) – between about 25 and 50 Gt of carbon from the atmosphere would have been sequestered. This accumulation of carbon in the forests is equivalent to between about 5 and 10 per cent of the emissions due to fossil fuel burning in the business-as-usual scenario during the first half of the next century. It would provide a useful contribution to any required reduction in atmospheric carbon dioxide concentrations.

But is such a tree planting programme feasible and is land on the scale required available? The answer is almost certainly, yes. Recent studies have identified land which is not presently being used for croplands or settlements, much of which has supported forests in the past, of an area totalling about 3.5 million square kilometres[5]. About 2.2 million square kilometres of this total is land which is technically suitable at mid and high latitudes – all of this is deemed to be available. In tropical regions, of the 22 million square kilometres actually deemed suitable, only 6 per cent or 1.3 million square kilometres is considered to be actually available because of additional cultural, social and economic constraints. These studies have also considered in detail how much carbon could be sequestered between the years 1995 and 2050 by a programme of forestation on this land. It is estimated to be between 50 and 70 Gt of carbon, to which a further 10–20 Gt can be added if the rate of tropical deforestation were to be slowed. Estimates of the cost of carrying out the programme have also emerged from the studies; they are considerably lower than those estimated earlier in the 1990s. When expressed per tonne of carbon sequestered they typically fall between 2 and 8 $(US) not including land and transaction costs, but also not including the value of local benefits (for instance, watershed protection, maintenance of biodiversity, education, tourism and recreation) which might be derived from the programme and which, in some circumstances, might offset most of the programme's cost. Compare this figure with the estimate given in Chapter 9 of between 50 and 100 $(US) for the cost per tonne of carbon of the likely damage due to global warming. The programme therefore appears as a potentially attractive one for alleviating the rate of change of climate due to increasing greenhouse gases in the relatively short term.

A possible afforestation programme has been presented in order to illustrate the potential for carbon sequestration. Once the trees are fully grown, of course, the sequestration ceases. What happens then depends on the use which may be made of them. They may be for 'protection' forests, for instance for the control of erosion or for the maintenance of biodiversity; or they may be production forests, used for biofuels or for industrial timber. If they are used for fuel, they add to the atmospheric carbon dioxide but, unlike fossil fuels, they are a renewable resource. As with the rest of the biosphere where natural recycling takes place on a wide variety of timescales, carbon from wood fuel can be continuously recycled through the biosphere and the atmosphere. As we shall see in the next chapter, biomass is seen as an important renewable energy source for the future.

Reduction in the sources of methane

Methane is a less important greenhouse gas than carbon dioxide, contributing perhaps 15 per cent to the present level of global warming. The stabilization of its atmospheric concentration would contribute a small but significant amount to the overall problem. Because of its much shorter lifetime in the atmosphere (about 12 years compared with 100–200 years for carbon dioxide), only a relatively small reduction in the anthropogenic emissions of this gas, about 8 per cent, would be required to stabilize its concentration at the current level.

Referring to Table 3.3 which lists the various sources of methane, there are three sources arising from human activities which could rather easily be reduced at small cost[6]. Firstly, methane emission from biomass burning would be cut by, say, one third if deforestation were drastically curtailed.

Secondly, methane production from landfill sites could be cut by at least a third if more waste were recycled or used for energy generation by incineration or if arrangements were made on landfill sites for the collection of methane gas (it could then be used for energy production or if the quantity were insufficient it could be flared, turning the methane into carbon dioxide which molecule for molecule is less effective than methane as a greenhouse gas). Waste management policies in many countries already include the encouragement of such measures.

Thirdly, the leakage from natural gas pipelines from mining and other parts of the petrochemical industry could at little cost (possibly even at a saving in cost) also be reduced by, say, one third. An illustration of the scale of the leakage is provided by the suggestion that the closing down of some Siberian pipelines because of the major recession in Russia has been the cause of the fall in the growth of methane concentration in the atmosphere from 1992 to 1993. Improved management of such installations could markedly reduce leakage to the atmosphere, perhaps by as much as one quarter overall.

Fourthly, with better management, options exist for reducing methane emissions from sources associated with agriculture[7].

Reductions from these four sources could reduce anthropogenic methane emissions by over 40 million tonnes per annum which would be more than adequate to stabilize the concentration of methane in the atmosphere at about or below the current level. Put another way, the reduction in methane emissions from these sources would be equivalent to a reduction in annual carbon dioxide emissions producing about one third of a gigatonne of carbon[8] or a little less than 5 per cent of total greenhouse gas emissions – a useful contribution towards the solution of the global warming problem.

Stabilization of carbon dioxide concentrations

We now turn to consider the stabilization of the atmospheric concentration of carbon dioxide, the most important of the greenhouse gases that result from human activities. Under a business-as-usual scenario, emissions will continue to increase throughout next century with no sign of abatement. The concentration of carbon dioxide also rises continuously through that period; it increases to about two and a half times its pre-industrial value by the year 2100 (Figs. 3.4 & 3.5). Under this sort of scenario of continually increasing emissions no stabilization of carbon dioxide concentration, and hence no stabilization of climate, is in sight.

What sort of emissions scenario would stabilize the carbon dioxide concentration? The commitment of some nations to ensure that carbon dioxide emissions in the year 2000 are no greater than they were in 1990 has already been mentioned. Suppose, after the year 2000, all countries kept their carbon dioxide emissions constant, would that be enough? Stabilizing concentrations is, however, very different from stabilizing emissions. With constant emissions after the year 2000, the concentration in the atmosphere would continue to rise and would approach 500 ppmv by the year 2100 (Fig. 3.5). After that the carbon cycle models predict that, because of the long time constants involved, the carbon dioxide concentration would still continue to increase although more slowly; there would be no stabilization for at least several hundred years.

An example of a scenario of emissions next century that would lead to stabilized carbon dioxide concentration is the WEC-C scenario presented in Figs. 3.4 and 3.5, in which carbon dioxide emissions are reduced well below current levels during the second half of the century. A much more comprehensive study with carbon cycle models, published by the IPCC, of emission profiles which would lead to the stabilization of carbon dioxide concentration at different levels (Fig. 10.1) demonstrates that this reduction in emissions below current levels is eventually required even for a stabilization level as high as 1000 ppmv, nearly four times the pre-industrial value.

In this study, many such pathways to stabilization could have been chosen. The particular pathways illustrated in Fig. 10.1 provide a smooth transition from the current average rate of carbon dioxide concentration increase to the time of stabilization; one set has been chosen to follow the emission scenario IS 92a for a longer period than the other. The profiles illustrate that, to a first approximation, the stabilized concentration level depends more on the

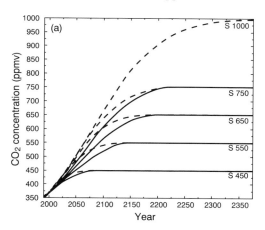
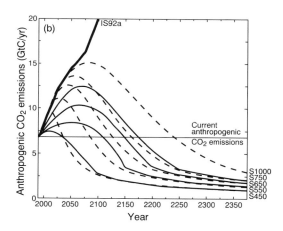

Fig. 10.1[9]
(a) carbon dioxide concentration profiles constructed so as to lead to stabilization at 450, 550, 650, 750 and 1000 ppmv. The dashed profiles allow emissions to follow the IS 92a scenario more closely during the early decades of the 21st century.
(b) carbon dioxide emissions leading to stabilisation of concentrations following the profiles in (a). The current anthropogenic carbon dioxide emissions and those for IS 92a are shown for comparison. To achieve eventual stabilization at 350 ppmv, emissions would need to be reduced to zero before the year 2050. The calculations use the Bern carbon cycle model and the carbon budget for the 1980s shown in Table 3.1.

Table 10.2
Total anthropogenic carbon dioxide emissions in Gt carbon accumulated from 1991–2100 inclusive; calculated using the carbon budget for the 1980s shown in Table 3.1 and the Bern carbon cycle model[10].

Case		Accumulated CO_2 emissions 1991 to 2100 (GtC)	
IS 92a scenario		1500	
Constant 1990 emissions		825	
		Concentration profiles A*	Concentration profiles B†
Stabilzation at	450 ppmv	630	650
	550 ppmv	870	990
	650 ppmv	1030	1190
	750 ppmv	1200	1300
	1000 ppmv	–	1410

* See Figure 10.1 (solid curves)
† Profiles that allow emissions to follow IS 92a until at least the year 2000 – see Figure 10.1 (dashed curves)

accumulated amount of carbon emitted up to the time of stabilization than on the exact concentration path followed en route to stabilization. This means that the alternative pathways which assume higher emissions in earlier years require steeper reductions in later years. Table 10.2 compares the accumulated emissions for the period 1991–2100 for the different pathways with those from the IS 92a scenario. It shows that if the atmospheric concentration of carbon dioxide is to remain below about 500 ppmv, the future global annual emissions averaged over the next century cannot exceed the current level of global annual emissions. Figure 10.2 shows the projected global mean surface temperature response to the carbon dioxide concentration profiles shown in Fig. 10.1a.

It is instructive also to look at annual emissions of carbon dioxide due to the combustion of fossil fuels expressed per capita[11]. Averaged over the world in 1990 they were about 1.1 tonnes (as carbon) but they varied very much from country to country. For developed countries and transitional economy countries in 1990 they averaged 2.8 tonnes (ranging from about 1.5 to 5.5 tonnes) while for developing countries they averaged about 0.5 tonnes (ranging from 0.1 to, in some few cases, over 2 tonnes). Looking ahead to the year 2100, if the world population rises to 12 billion (the United Nations base case), under the profiles of carbon dioxide emissions leading to stabilization at concentrations of 450 ppmv, 550 ppmv and 650 ppmv respectively (Fig. 10.1) the per capita annual emissions averaged over the world would be about 0.25, 0.5 and 0.7 tonnes – all much less than the current value of 1.1 tonnes.

Realizing the Climate Convention Objective

The last few sections have addressed the main greenhouse gases and how their concentrations might be stabilized. To decide how the appropriate stabilization

Fig 10.2
Projected temperature response to the carbon dioxide concentration profiles shown in Fig 10.1a. A temperature sensitivity for equilibrium doubled carbon dioxide of 2.5 °C is assumed The calculations assume the 'observed' history of forcing to 1990, including aerosol effects, but then carbon dioxide changes only after 1990[12]. To allow for the temperature rise from pre-industrial times to the present, add about 0.5 °C to the curves.

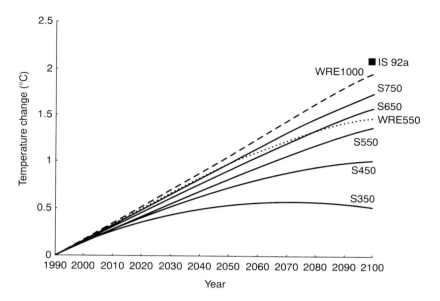

levels should be chosen as targets for the future we look to the guidance provided by the Climate Convention Objective (see box on p. 144) which states that the levels and the timescales for their achievement should be such that dangerous interference with the climate system must be prevented, that ecosystems should be able to adapt naturally, that food production must not be threatened and that economic development can proceed in a sustainable manner. We do not yet know enough to pick precisely the levels or the timescales under the criteria the Climate Convention is prescribing, but perhaps already some limits can be set.

Firstly, considering the most important greenhouse gas, carbon dioxide, as we have already noted, its long life in the atmosphere provides severe constraints on the future emission profiles which lead to stabilization at any level. It will be clear, for instance, from Fig. 10.1 that stabilization below about 400 ppm would require an almost immediate drastic reduction in emissions. Such reduction could only be achieved at a large cost and with some curtailment of energy availability and would almost certainly breach the criterion which requires 'that economic development can proceed in a sustainable manner'.

What about the upper end of the choice of level? Here we refer to the likely impacts of climate change under a situation in which the atmospheric concentration of carbon dioxide has doubled from its pre-industrial value of 280 ppmv to about 560 ppmv. Many of the impacts described in Chapter 7 with their associated costs apply to this situation. We noted in Chapter 9 that, as far as we are able to estimate at the moment, the costs of the likely damage of the impacts at that level of climate change were larger than the costs of stabilizing carbon dioxide concentration at levels above 400 or 450 ppmv (Fig. 9.4). We also noted that, beyond the doubled carbon dioxide situation, the likely damage due to greenhouse gas climate change is likely to rise substantially more rapidly as the amount of carbon dioxide in the atmosphere increases. A further factor is the rate of climate change (see Fig. 10.2) which, with all the profiles in Fig. 10.1 except possibly the two lowest, is likely to be such that some important ecosystems may not be able to adapt to it (see Chapter 7). Considering carbon dioxide alone, these considerations suggest that the range between about 400 ppmv and 550 ppmv is where further careful consideration of the choice of the target stabilization level should be made.

Although carbon dioxide is the most important greenhouse gas, other gases also make a contribution to climate change. As we saw in Chapter 3 the combined effect of the increases to 1990 of the gases methane, nitrous oxide and the CFCs[13] is to add a forcing equivalent to that from an additional 45 ppmv or so of carbon dioxide. The effect of these other gases also needs to be taken into account in our overall discussion of the Climate Convention Objective of stabilization. If there were no further increase in these minor gases, the 1990 forcing would still require to be added to future projections of change. The effect of this, if turned into equivalent amounts of carbon dioxide, would be that the 450 ppmv carbon dioxide only level would become just over 500 ppmv and the 550 ppmv level would become about 620 ppmv of equivalent carbon dioxide[14]. This means that, if it is considered that the climate effects of doubled pre-industrial carbon dioxide concentration should be an upper limit, when the increases in other gases are allowed for, the stabilization limit for carbon dioxide only is no more than 500 ppmv.

How realistic is it to assume that the concentration of the other gases will not change? We saw earlier that the Montreal Protocol should ensure that the CFCs are stabilized in concentration over the next decade or two. We also saw, for methane, that means are available which are not costly and which, if taken, could stabilize methane concentrations at about today's levels. There is more uncertainty about nitrous oxide as its sources and sinks are not well known. However, it is only a small contributor to the forcing to date (equivalent to about 10 ppmv of carbon dioxide); any increase in the future is not likely to have a large effect.

In this simple argument regarding the influence of other gases on the choice of a concentration level for carbon dioxide stabilization which might be acceptable under the terms of the Climate Convention Objective, the concept of equivalent carbon dioxide concentration has proved a useful tool. But it must not be used blindly. For any detailed consideration of the choice of level, there are other scientific factors to be included. Firstly, there are the other contributors to radiative forcing and climate change namely tropospheric ozone and aerosols which are very inhomogeneous in their distribution; their likely effect, although small compared with that of carbon dioxide, also needs to be taken into account[15]. Secondly, there are the different regional climate responses and different timescales of responses which result from the different greenhouse gases or from aerosols. Thirdly, there are the effects of particular feedbacks (eg carbon dioxide fertilisation) or impacts (eg acid rain from aerosols).

The choice of a target stabilization level for greenhouse gases according to the criteria listed in the Objective of the Climate Convention involves scientific, economic, social and political factors. Within these latter factors considerations of equity are central to the discussion. In Chapter 9, weighing the importance of the needs of the present generation against those of future generations – or inter-generational equity – was mentioned as a core component of the concept of sustainable development. Also of importance is the balance of equity between the industrial and developed nations and the developing world – we can call it international equity. This balance is recognized in the Framework Convention on Climate Change (see box) where the growing energy needs of developing nations as they achieve industrial development are clearly stated.

Taking all factors into consideration will involve different kinds of analysis, cost benefit analysis (which we considered briefly in Chapter 9), multicriteria analysis (which takes into account factors which cannot be expressed in monetary terms) and sustainability analysis (which considers avoidance of particular thresholds of stress or of damage). Further, because much uncertainty is associated both with many of the factors which have to be included and with the methods of analysis, the process of choice is bound to be an evolving one subject to continuous review – a process often described as sequential decision making.

Finally, we look at an example of a pathway to the stabilization next century of carbon dioxide concentration at about 450 ppmv – within the range we have mentioned above – as provided by the World Energy Council ecologically driven scenario (Scenario C) described in Chapter 3 (Figs. 3.4 & 3.5). Under that scenario, global carbon dioxide emissions grow by about 10 per cent by the year 2050; they then fall by a factor of two by the end of next century. For the first two decades of next century, the World Energy Council provide detailed projections for Scenario C which recognize the requirement for international equity. Up to the year 2020, emissions from fossil fuels in the developing world are allowed to approximately double, while those from developed countries fall by about 30 per cent. In 2020, global emissions from developing countries would be 60 per cent of the total for the world compared with about one third in 1990. After 2020, it is clear that if the scenario is to be realized, the growth in emissions in developing countries cannot continue to be maintained. For example, the scenario's target of total emissions from the world in 2100 – about half of those in 1990 – is approximately equal to the projected emissions from developing countries alone in the year 2020. Achievement of that scenario will therefore require reductions in emissions by all nations during the greater part of the next century.

As the World Energy Council point out in their report, achievement of such a scenario will be far from easy. It requires three essential ingredients. The first is an aggressive emphasis on energy saving and conservation. Much here can be achieved at zero net cost or even at a cost saving. Though much energy conservation can be shown to be economically advantageous, it is unlikely to be undertaken without significant incentives. However, it is clearly good in its own right, it can be started in earnest now and it can make a significant contribution to the reduction of emissions and the slowing of global warming. The second ingredient is an emphasis on the development of appropriate non-fossil fuel energy sources leading to very rapid growth in their implementation. The third is the transfer of technologies to developing countries which will enable them to apply the most appropriate and the most efficient technologies to their industrial development, especially in the energy sector.

Summary of the action required

This chapter has suggested some actions that can be taken to slow climate change and ultimately to stabilize it as required by the internationally agreed Climate Convention.

Some actions have already been taken which have an effect on global emissions of greenhouse gases, namely:

- the reduction by some countries of carbon dioxide emissions in the year 2000 to 1990 levels and
- the provisions of the Montreal Protocol regarding the emissions of CFCs and CFC substitutes.

Other actions which can be taken now to slow climate change, which can be done at little or no net cost and which are good to do for other reasons are the following:

- a reduction of deforestation,
- a substantial increase in afforestation,
- some easy-to-do reductions in methane emissions,
- an aggressive increase in energy saving and conservation measures,
- increased implementation of renewable sources of energy supply.

For the longer term, as well as increased emphasis on these actions, the world needs to begin to follow an energy scenario which will lead to the stabilization of carbon dioxide concentration in the atmosphere. The choice of a target stabilization level following the guidance of the Climate Convention involves the consideration of many factors and, because of the uncertainties, will necessarily be subject to continuous review. We have presented arguments suggesting that, at the current state of knowledge, the range 400–500 ppmv in carbon dioxide concentration is where further detailed consideration of costs and impacts should be concentrated. An example of an energy scenario which would stabilize carbon dioxide concentrations in this range is the World Energy Council scenario C. Its realization will require very rapid growth in the implementation of appropriate non-fossil fuel energy sources; it will also require that means be provided to enable developing countries to apply appropriate and efficient technologies to their industrial development, especially in the energy sector – matters which will be addressed in detail in the next chapter.

Questions   1   From Fig. 10.2, what are the rates of change of global average temperature for the profiles shown in Fig. 10.1 which lead to stabilization of carbon dioxide concentration at different levels? From information in Chapter 7 or from elsewhere, can you suggest a criterion involving rate of change which might assist in the choice of a stabilization level for carbon dioxide concentration as required by the Objective of the Climate Convention.

2   From the formula in Note 14 and the information in Fig. 3.9, calculate the contributions from the various components of radiative forcing (including aerosol)

to the equivalent carbon dioxide concentration in 1990. How valid do you think is it to speak of equivalent carbon dioxide for components such as aerosol and tropospheric ozone?

3 Associated with the choice of stabilization level under the criteria of the Objective of the Climate Convention, different kinds of analysis were mentioned; cost benefit analysis, multicriteria analysis and sustainability analysis. Discuss which analysis is most applicable to each of the criteria in the Objective. Suggest how the analyses might be presented together so as to assist in the overall choice.

4 From the information available in previous chapters and using the criteria laid out in the Climate Convention Objective, what stabilization levels of greenhouse gas concentrations do you think should be chosen?

5 The arguments concerning the choice of stabilization level and the action to be taken have concentrated on the likely costs and impacts of climate change before the year 2100. Do you think that information about continuing climate change or sea-level rise (see Chapter 7) after 2100 should be included and taken into account by decision makers, or is that too far ahead to be of importance?

6 The international response to Global Warming is likely to lead to decisions being taken sequentially over a number of years as knowledge regarding the science, the likely impacts and the possible responses becomes more certain. Describe how you think the international response might progress over the next twenty years. What decisions might be taken at what time?

7 Using projections of population statistics for developed and developing countries, suggest how per capita emissions might be shared between developed and developing countries in the years 2050 and 2100 if emissions of carbon dioxide are constrained to follow the profiles leading to stabilization at 450, 550 and 650 ppmv.

8 Find out the details of any plans for afforestation in your country? What actions or incentives could make it more effective?

Notes

1 More detail in *The World Environment 1972–1992*, eds. M. K. Tolba, O. A. El-Kholy, Chapman and Hall, 1992, pp. 157–82.

2 Quoted in *The World Environment 1972–1992*, eds. M. K. Tolba, O. A. El–Kholy, Chapman and Hall, 1992, p. 169.

3 These figures do not include carbon in the soils. See Salati *et al.*, in *Climate Change: Science, Impacts and Policy; Proceedings of the Second World Climate Conference*, eds. J. Jager and H. L. Ferguson, CUP, 1991, pp. 391–5; also J. Leggett *et al.*, 'Emission scenarios for the IPCC: an update in climate change', IPCC, 1992, p. 89; a more detailed source is R. A. Houghton, *Climate Change*, **19**, 1991, pp. 99–118.

4 S. Brown *et al.*, 'Management of forests for mitigation of greenhouse gas emissions', chapter 24 in *Climate Change 1995: Impacts, Adaptations and Mitigation of Climate Change*, eds. R. T. Watson, M. C. Zinyowera and R. H. Moss, CUP, 1996.

5 S. Brown *et al.*, 'Management of forests for mitigation of greenhouse gas emissions', chapter 24 in *Climate Change 1995: Impacts, Adaptations and Mitigation of*

Climate Change, eds. R. T. Watson, M. C. Zinyowera and R. H. Moss, CUP, 1996.

6   J. C. Hourcade *et al.*, 'A review of mitigation cost studies', chapter 9 in *Climate Change 1995: Economic and Social Dimensions of Climate Change*, eds. J. Bruce, Hoesung Lee and E. Haites, CUP, 1996.

7   V. Cole *et al.*, 'Agricultural options for mitigation of greenhouse gas emissions', chapter 23 in *Climate Change 1995: Impacts, Adaptations and Mitigation of Climate Change*, eds. R. T. Watson, M. C. Zinyowera and R. H. Moss, CUP, 1996.

8   This figure is calculated by multiplying the 60 million tonnes by the global warming potential for methane which, for a time horizon of 100 years, is about 7.5 (molecule CH_4 per molecule CO_2 – see note 17 in Chapter 3), then by 12/16 to put it into tonnes of C.

9   From Technical Summary *Climate Change 1995: The Science of Climate Change*, eds. J. T. Houghton, L. G. Meira Filho, B. A. Callender, N. Harris, A. Kattenberg and K. Maskell, CUP, 1996: Information regarding 350 ppm curves is available from *IPCC Technical Paper No 3*, D. Schimel *et al.* 'Stabilisation of atmospheric greenhouse gases: physical, biological and socio-economic implications' IPCC, Geneva, 1997.

10 Adapted from Table 5 in the Technical Summary, *Climate Change 1995: The Science of Climate Change*, eds. J. T. Houghton, L. G. Meira Filho, B. A. Callender, N. Harris, A. Kattenberg and K. Maskell, CUP, 1996.

11 *IPCC Second Assessment* synthesis of scientific-technical information relevant to interpreting Article 2 of the UN Framework Convention on Climate Change, IPCC, Geneva 1995.

12 From *IPCC Technical Paper No 3*, D. Schimel *et al.* 'Stabilisation of atmospheric greenhouse gases: physical, biological and socio-economic implications' IPCC, Geneva, 1997.

13 Allowing for the CFCs a reduction in their forcing because of stratospheric ozone destruction.

14 Note that, although the amount of forcing from the minor gases is the same, when turned into equivalent carbon dioxide, the amounts added increase with the carbon dioxide concentration to which the amount is added. This is because the relationship between radiative forcing (R in Wm^{-2}) and concentration (C in ppm) is non-linear. The relationship is $R = 6.3 \ln (C/C_0)$ where C_0 is the pre-industrial CO_2 concentration.

15 More detail in *IPCC Technical Paper No 3*, D. Schimel *et al.* 'Stabilisation of atmospheric greenhouse gases: physical, biological and socio-economic implications' IPCC, Geneva, 1997.

11 Energy and Transport for the Future

We flick a switch and energy flows. Energy is provided so easily for the developed world that thought is rarely given to where it comes from, whether it will ever run out or whether it is harming the environment. Energy is also cheap enough that little serious attention is given to conserving it. However, most of the world's energy comes from the burning of fossil fuels which generates a large proportion of the greenhouse gas emissions into the atmosphere If these emissions are to be reduced, a large proportion of the reduction will have to occur in the energy sector. There is a need, therefore, to concentrate the minds of policymakers and indeed of everyone on our energy requirements and usage. This chapter looks at how future energy might be provided in a sustainable manner.[1]

World energy demand and supply

Most of the energy we use can be traced back to the sun. In the case of fossil fuels it has been stored away over millions of years in the past. If wood (or other biomass including animal and vegetable oils), hydro-power, wind or solar energy itself is used, the energy has either been converted from sunlight almost immediately or has been stored for at most a few years. These latter sources of energy are renewable; they will be considered in more detail later in the chapter. The only common form of energy that does not originate with the sun is nuclear energy; this comes from radioactive elements which were present in the Earth when it was formed.

Until the Industrial Revolution, energy for human society was provided from 'traditional' sources – wood and other biomass and animal power. Since 1860, as industry has developed, the rate of energy use has multiplied by about a factor of 30 (Fig. 11.1), at first mostly through the use of coal followed, since about 1950, by rapidly increasing use of oil. In 1990 the world consumption of energy was about 8,730 million tonnes of oil equivalent (toe). This can be converted into physical energy units to give an average rate of energy use of about 12 million million watts (or 12 Terawatts $= 12 \times 10^{12}$ W)[2].

Great disparities exist in the amount of energy used per person in various parts of the world (Fig. 11.2). In 1990 each person in the world used on average 1.65 toe, an average consumption of energy of about 2.2 kilowatts (kW). The highest rates of energy consumption are in North America where the average citizen in 1990 used nearly 8 toe, equivalent to an average rate of consumption

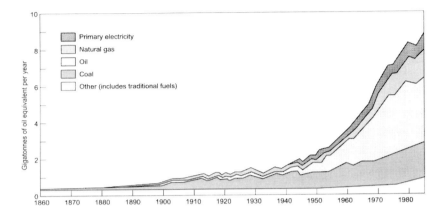

Fig. 11.1
Growth in the rate of energy use and in the sources of energy since 1860 in thousand
millions of tonnes of oil equivalent (Gtoe) per year[3]. In terms of primary energy units,
1 Gtoe = 41.87 exajoules (1 exajoule (EJ) $=10^{18}$J). The energy shown for electricity is
that used in its generation (for which the basic unit is the watt-hour(thermal)); the
efficiency of conversion into usable electrical energy (for which the basic unit is the
watt-hour (electrical)) is typically around one third.

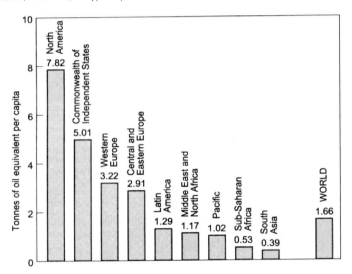

Fig. 11.2
Per capita annual energy use in tonnes of oil equivalent in 1990 in different regions of
the world[4].

of about 11 kW. By contrast the average Indian used only one twentieth of that
amount or about 0.4 toe, equivalent to an average rate of consumption of 0.5
kW, mainly in the form of traditional fuels. In fact about half the world's population rely wholly on traditional fuels and do not currently have access to commercial energy in any of its forms.

It is interesting to see how the energy we consume is used. Taking the world
average, about 20 per cent of primary energy is used in transportation, about 45

Table 11.1
Reserves of fossil fuels and their relation to current energy use. Data from World Energy Council[5].

	Proven reserves in 1990 (Gtoe)	Ratio: Reserves to current annual use (years)	Ultimately recoverable resources (Gtoe)
Coal (excluding lignite)	496	197	} 3400
Lignite	110	293	
Conventional oil	137	40	200
Unconventional oils (heavy crude, natural bitumen, oil shale)			600
Natural gas	108	56	220

per cent by industry and the remaining 35 per cent in commercial activity and in homes. It is also perhaps interesting to know how much energy is used in the form of electricity. About one third of primary energy goes to make electricity at an average efficiency of conversion of about one third. Of this electrical power about half, on average, is utilized by industry and the other half in commercial activities and in homes.

How much is spent on energy? Taking the world as a whole, the amount spent per year by the average person for the 1.65 toe of energy used, is about 5 per cent of annual income. Despite the very large disparity in incomes, the proportion spent on primary energy is much the same in developed countries and developing ones.

How about energy for the future? If we continue to generate most of our energy from coal, oil and gas, do we have enough to keep us going? Current knowledge of proven recoverable reserves (Table 11.1) indicates that known reserves of fossil fuel will meet demand for the period up to 2020 and substantially beyond. Around mid-century, if demand continues to expand, oil and gas production will come under increasing pressure. Further exploration will be stimulated which will lead to the exploitation of more sources, although increased difficulty of extraction can be expected to lead to a rise in price. So far as coal in concerned, there are operating mines with resources for production for well over a hundred years.

Estimates have also been made of the ultimately recoverable fossil fuel reserves, defined as those potentially recoverable assuming high but not prohibitive prices and no significant bans on exploitation. Although these are bound to be somewhat speculative[6], they show (Table 11.1) that, at current rates of use, reserves of oil and gas are likely to be available for 100 years and of coal for more than 1,000 years. In addition to fossil fuel reserves considered now to be potentially recoverable there are reserves not included in Table 11.1, such as the methane hydrates, which are probably very large in quantity but from which extraction would be much more difficult.

Likely reserves of uranium for nuclear power stations should also be included in this list. When converted to the same units (assuming their use in 'fast' reactors) they are believed to be at least 3,000 and possibly as high as 9,000 Gtoe, substantially greater than likely fossil fuel reserves.

For at least the next century, sufficient fossil fuels in total are available to meet likely energy demand. It is considerations other than availability, in particular environmental considerations, which will tend to limit fossil fuel use.

Future energy projections

Many energy experts, individuals and organizations have attempted to look into the crystal ball of the future and estimate future energy demand next century and how it might be met. Any such attempts are bound to be fraught with a lot of uncertainty and to include assumptions of dubious accuracy; no one really knows how human beings will behave or what political, institutional or technical changes may occur. Nevertheless a range of possibilities must be considered in order to predict what climate change might occur. The predictions of climate change drawn up in Chapter 6 based on a business-as-usual scenario prepared under the auspices of the IPCC (scenario IS 92a) estimated future energy demand on the assumption that no controls or constraints for environmental reasons were applied. It is interesting to compare the energy assumptions of that scenario with four energy scenarios constructed by the World Energy Council (WEC)[7]. These scenarios (more details in the box) take into account likely population growth and energy sources and a realistic view of the rate of technical change. One of the scenarios, the 'high growth' case (A), assumes a higher rate of economic growth in developing countries. An 'ecologically driven' case (C) assumes that environmental pressures have a large influence on energy demand and growth. The other scenarios (B and B1), denoted reference cases, are based on moderate assumptions about economic growth.

Details of the scenarios to the year 2020 are shown in Figs. 11.3 and 11.4. As can be seen from Fig. 11.3, it is only in the developed world that there is potential for containing future energy demand. Population growth and the need for economic development in developing countries make it inevitable that they will, for many decades, consume increased amounts of energy. For all the scenarios to 2020, fossil fuels continue to dominate the energy mix (Fig. 11.4). The contribution from nuclear power is assumed to grow in all the cases. New renewable energy sources play an increasing role, although apart from Case C their contribution is modest.

For the 'ecologically driven' WEC scenario C the energy demand in 2020 is about 30 per cent more than that in 1990 and 30 per cent less than that for scenario A. That scenario C assumes both that there will be large increases in efficiency leading to a reduced energy demand and also, following the results of a WEC study on renewable energy[8], a substantial growth in the share of primary energy supply coming from new renewable energy sources ('modern' biomass, solar, wind and so on). A growth in energy supply from these new renewable sources from 2 per cent in 1990 to 12 per cent in 2020 (by when 1.4 Gtoe per year would be coming from these sources) is considered feasible if their development is given sufficient support. By the year 2100 under scenario C, 50 per cent of energy supply is assumed to come from these sources. The WEC report points out that 'cost effective research, development and installation involving

World Energy Council and IPCC scenarios

The World Energy Council (WEC) is an international non-governmental organization with representation from all parts of the energy industry and from over 90 countries. The Council has developed four energy scenarios for the period to 2020, each representing different assumptions in terms of economic development, energy efficiencies, technology transfer and the financing of development round the world[9]. The WEC emphasizes that they have been developed to illustrate future possibilities and they should not be considered as predictions. For all four cases, following the United Nations base case, world population is assumed to grow from 5.3 thousand million in 1990 to 8.1 thousand million in 2020 (and further to 10 thousand million in 2050 and 12 thousand million in 2100), with more than 90 per cent of this growth in the developing world. These are similar figures to those produced by the World Bank in 1991 which were assumed in the IPCC IS 92a scenario.

All four cases assume that there will be significant environmental and economic pressures to achieve major improvements in energy efficiency compared to historic performance, although to different degrees within the various economic groupings of countries. One of them, scenario C, assumes very strong pressure to reduce the emissions of greenhouse gases in order to combat global warming. Table 11.2 presents the detailed assumptions underlying the four scenarios.

Table 11.2 refers to the 'energy intensity' which is the ratio of energy use to Gross Domestic Product (GDP); it is a measure of energy efficiency. When averaged over the world, over the past fifty years it has been falling by about 1 per cent per year. A more demanding rate of reduction in energy intensity than this 1 per cent is assumed for all the scenarios; for case C, the ecologically driven scenario, the rate assumed is considered very demanding indeed. The main difference between the modified reference case B1 and the reference case B is that, in B1, the rate of reduction of energy intensity assumed for the economies in transition is less than in case B

Table 11.2
Assumptions underlying the four WEC energy scenarios[10]. See glossary for explanation of abbreviations.

Case	A	B1	B	C
Name	High growth	Modified reference	Reference	Ecologically driven
Economic growth % p.a.	High	Moderate	Moderate	Moderate
OECD	2.4	2.4	2.4	2.4
CEE/CIS	2.4	2.4	2.4	2.4
DCs	5.6	4.6	4.6	4.6
World	3.8	3.3	3.3	3.3
Energy intensity reduction % p.a.	High	Moderate	High	Very high
OECD	−1.8	−1.9	−1.9	−2.8
CEE/CIS	−1.7	−1.2	−2.1	−2.1
DCs	−1.3	−0.8	−1.7	−2.4
World	−1.6	−1.3	−1.9	−2.4
Technology transfer	High	Moderate	High	Very high
Institutional improvements (world)	High	Moderate	High	Very high
Possible total demand (Gtoe)	Very high	High	Moderate	Low
	17.2	16.0	13.4	11.3

World Energy Council and IPCC scenarios *continued*

and for the developing countries is only half that in case B.

With somewhat less detail, scenarios A, B and C have been extended to the year 2100. Global energy demand can be expected to continue to increase, but by that time the availability of fossil fuels will be more limited and new renewables can be expected to contribute substantially to the energy mix for all scenarios. Some of the characteristics of the scenarios out to 2100 are listed in Table 11.3 – similar details of the scenarios for 2020 are in Fig 11.4.

The IPCC scenario IS 92a – the business-as-usual scenario – assumes moderate economic growth and no strong environmental pressures to cut emissions (more details in the caption to Table 11.3).

Table 11.3
Some characteristics of the WEC scenarios out to the year 2100[11]. For Scenario IS 92a, the comparable figures are: 40 Gtoe for the global energy demand in 2100 (about 53% supplied by fossil fuels) and 13.2 and 19.8 Gt respectively for the CO_2 emissions from fossil fuels in 2050 and 2100.

Case	1990	A		B		C	
		2050	2100	2050	2100	2050	2100
Global energy demand (Gtoe)	8.8	27	42	23	33	15	20
Fossil fuels (% of primary energy)	77	58	40	57	33	58	15
Nuclear (% of primary energy)	5	14	29	15	28	8	11
New renewables (% of primary energy)	2	15	24	14	26	20	50
Annual CO_2 emissions from fossil fuels (GT carbon)	6.0	14.9	16.6	12.2	11.7	7.3	2.5
Annual CO_2 emissions from fossil fuels (% change on 1990)		152	181	107	98	24	−59

financing which only governments can supply will be needed if these sources of energy are to be implemented on the large scale shown in the Ecologically Driven Case C'. Renewable energy sources will be discussed further later in the chapter.

It is interesting to compare the future projections generated by the WEC with the IPCC business-as-usual scenario IS 92a. Up to the year 2020 the energy demand and the carbon dioxide emissions (Fig. 3.4) for IS 92a are similar to those of the WEC reference scenario B1. After about the middle of the century, the WEC scenarios assume more strongly than the IPCC ones that growth in energy demand and also in carbon dioxide emissions will begin to be limited by the increasing scarcity and price of oil and gas.

For all these scenarios, except WEC scenario C, atmospheric concentrations of carbon dioxide (Fig. 3.5) continue to rise throughout next century. The WEC scenario C comes near to stabilizing emissions over the period to 2020 and to stabilizing atmospheric concentrations by 2100 (see Figs. 3.4 and 3.5).

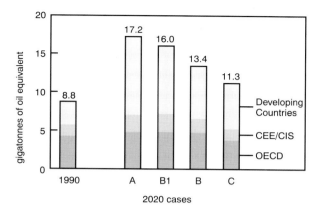

Fig. 11.3
Primary energy demand by economic groupings of countries for WEC scenarios[12].

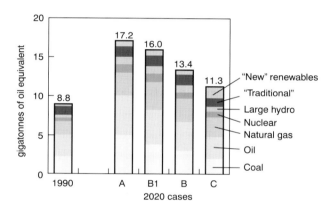

Fig. 11.4
Primary energy supply mix for WEC scenarios[13].

In its 1995 Report, the IPCC has developed a number of projections (the Low-Emissions Supply Systems or LESS constructions[14]) showing how the necessary energy supply next century could be provided with different options of energy supply especially with respect to renewable energies. Some details of these will be presented later in the chapter (see box on p. 220).

The following sections address how increased energy conservation and efficiency can be achieved and what developments can be realized in new renewable energy sources; these are the technical means through which the necessary reductions in carbon dioxide emissions will be achieved in the energy sector.

Energy conservation and efficiency

If we turn lights off in our homes when we do not need them, if we turn down the thermostat by a degree or two so that we are less warm or if we add more insulation to our home, we are conserving or indeed saving energy. But are such actions significant in overall energy terms? Is it realistic to plan for really worthwhile savings in our use of energy?

To illustrate what might be possible, I will consider the efficiency with which energy is currently used. The energy available in the coal, oil, gas, uranium, hydraulic or wind power is *primary energy*. It is either used directly, for instance as heat or it is transformed into motor power or electricity which in turn provides for many uses. The process of energy conversion, transmission and transformation into its final useful form involves a proportion of the primary energy being wasted. For example, to provide 1 unit of electrical power at the point of use typically requires about 3 units of primary energy. An incandescent light bulb is about 3 per cent efficient in converting primary energy into light energy; unnecessary use of lighting reduces the overall efficiency to perhaps no more than 1 per cent[15]. Assessments have been carried out across all energy uses comparing actual energy use with that which would be consumed by ideal devices providing the same services. Although there is some difficulty in defining precisely the performance of such 'ideal' devices (see box for a discussion of thermodynamic efficiencies), assessments of this kind come up with world average end-use energy efficiencies of the order of 3 per cent. That sort of figure suggests that there is a large amount of room for improvement in energy efficiency, perhaps by at least threefold[16]. The following paragraphs consider three areas where there are real possibilities of savings: in buildings, in industry and in transport.

Thermodynamic efficiencies

When considering the efficiency of energy use, it can be important to distinguish between efficiency as defined by the First Law of Thermodynamics and efficiency as defined by the Second Law. The second particularly applies when energy is used for heating.

A furnace used to heat a building may deliver to heating the building say 80 per cent of the energy released by full combustion of the fuel, the rest being lost through the pipes, flue etc. That 80 per cent is a First Law efficiency. An ideal thermodynamic device delivering 100 units of energy as heat to the inside of a building at a temperature of 20 °C from the outside at a temperature of 0 °C would only require just under 7 units of energy. So the Second Law efficiency of the furnace is less than 6 per cent.

Heat pumps (refrigerators or air conditioners working in reverse) are devices which make use of the Second Law and deliver more energy as heat than the electrical energy they use. Typically their Second Law efficiencies are about 30 per cent, which means that they can still deliver more heat energy than the primary energy required to generate the electricity they use. Because of their comparatively high capital and maintenance costs, however, heat pumps have not been widely used. An example of their substantial use is their contribution to district heating in the city of Uppsala in Sweden where 4 MW of electricity is employed to extract heat from the river and deliver 14 MW of heat energy.

First of all, what about energy in buildings? So that we are comfortable in them we heat them in winter and we cool them in summer. In the United States, for instance, about 36 per cent of the total use of energy is in buildings (about two-thirds of this in electricity), including about 20 per cent for their heating (including water heating) and about 3 per cent for cooling them[17].

Two ways in which really substantial energy savings can be made in buildings are by improving their insulation (see box on p. 198) and by improving the efficiency of appliances (see box). Many countries, including the UK and the USA still have relatively poor standards of building insulation compared, for instance, with Scandanavian countries. Improvements in building design to make better use of energy from sunlight can also help (see box on p. 211). There are also large possibilities for the improvement of the efficiency of appliances at relatively small cost.

The results of a study in the United States have identified some of the large savings which could be made in the electricity used in buildings. The cost of such action would be less than the cost of the energy savings; overall therefore there would be a substantial net saving (Fig. 11.5). The twelve options in

Efficiency of appliances

In Table 11.4, figures are given for the average annual electricity consumption of typical domestic appliances used in a home in the United Kingdom. If, in replacing appliances, everyone bought the most efficient available, their total electricity consumption could easily drop by more than half.

Savings in electricity also deliver a net economic return to the user. Take for example fluorescent light bulbs which are as bright as ordinary light bulbs, but use a quarter of the electricity and last eight times as long before they have to be replaced. A 15 watt compact fluorescent bulb (equivalent to a 60 watt ordinary incandescent light bulb) costing £10 will use about £10 worth of electricity over its lifetime. To cover the same period eight ordinary bulbs would be needed costing about £4 but using £40 worth of electricity. The net saving is therefore about £24. Similar calculations could be carried out for the other appliances.

Table 11.4
Annual appliance electricity consumption in kWh[18].

(kWh/year)	1990 Existing stock (UK)	Average new model (UK)	Best available model (worldwide)
Cooker	840	780	370
Washing machine	210	180	70
Dishwasher	500	430	300
Refrigerator	350	300	60
Fridge/Freezer	730	500	275
TV	200	140	100
Lighting	370	370	105

Insulation of buildings

About one and a half thousand million people live in cold climates where some heating in buildings is required. In most countries the energy demand of space heating in buildings is far greater than it need be if the buildings were better insulated.

Table 11.5 provides as an example details of two houses, showing that the provision of insulation in the roof, the walls and the windows can easily lead to the energy requirement for space heating being more than halved (from 5.8 kW to 2.65 kW). The cost of the insulation is small and is quickly recovered through the lower energy cost.

If a system for circulating air through the house is also installed, the number of necessary air changes with outside air is less and the total heating requirement further reduced. In this case it is worthwhile to add more insulation to reduce the heating requirement still further.

Table 11.5
Two assumptions (one poorly insulated, and one moderately well insulated) regarding construction of a detached, two-storey house with ground floor of size 8 m × 8 m, and the accompanying heat losses (U-values express the heat conduction of different components in watts per square metre per °C.)

	Poorly insulated	Moderately well insulated
Walls 150 m^2 total area	Brick + cavity + block U-value 0.7	Brick + cavity + block with insulation in cavity of 75 mm thickness: U-value 0.3
Roof 85 m^2 area	Uninsulated U-value 2.0	Covered with insulation of thickness 150 mm: U-value 0.2
Floor 64 m^2	Uninsulated U-value 1.0	Includes insulation of thickness 50 mm: U-value 0.3
Windows 12 m^2 total area	Single glazing U-value 5.7	Double glazing with low emissivity coating: U-value 2.0
Heat losses (in kW) with 10 °C temperature difference from inside to outside	roof 1.7 walls 1.1 windows 0.7 floor 0.7	roof 0.2 walls 0.45 windows 0.2 floor 0.2
Total heat loss (kilowatts)	4.2	1.05
Add heat (in kilowatts) needed for air changes (1.5 per hour)	1.60	1.60
Total heating required (kilowatts)	5.8	2.65

Fig. 11.5 together cover about 45 per cent of the amount of electricity used in residential buildings in the USA, which in 1989 was 1630 TWh or about 10 per cent of the USA's total energy use. The four options which provide the largest savings (together adding up to 60 per cent of the savings) are in the areas of commercial lighting, commercial air conditioning, residential appliances and residential space heating. Electricity companies in some parts of the United States are contracting to implement some of these energy saving measures as an alternative to the installation of new capacity – at significant profit both to the companies and its customers. Similar savings would be possible in other developed countries. Major savings at least as large in percentage terms could also be made in countries with economies in transition and in developing countries if existing plant and equipment were used more efficiently.

Similar room for efficiency savings exists in industry. The installation of relatively simple control technology often provides large potential for energy reduction at a substantial net saving in cost. The cogeneration of heat and power, which already enables electricity generators to make better use of heat which would otherwise be wasted, is particularly applicable to some industrial plants where large amounts of both heat and power can be required. To take an example: British Sugar with an annual turnover of £700 million now spends

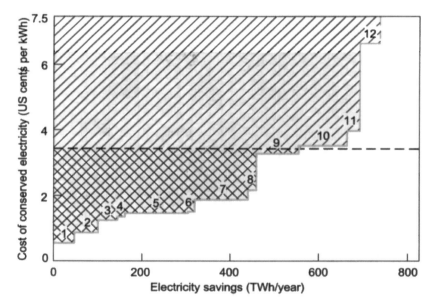

Fig. 11.5
Cost of various options (at 1989 prices) for saving electricity in buildings[19]. If the cost of conservation is less than the cost of the electricity saved over the life-time of the application, a net saving results. The various options are: (1) use of white surfaces to reduce the need for air conditioning; (2) residential lighting; (3) residential water heating; (4) commercial water heating; (5) commercial lighting; (6) commercial cooking; (7) commercial cooling; (8) commercial refrigeration; (9) residential appliances; (10) residential space heating; (11) commercial and industrial space heating; (12) commercial ventilation. The shaded areas are all below 7.5 cents per kWh (the all-sector average electricity price) and 3.5 cents per kWh (typical operating cost of US electricity generation).

The growth of motor transport

Since the 1950s there has been a phenomenal growth in road transport (Fig. 11.6) especially in the industrialized Western countries. In these countries road transport accounts for over 80 per cent of all transport and also for nearly 50 per cent of total oil consumption.

The potential for future growth is not only illustrated by the trend shown in Fig. 11.6, but also by the enormous disparity in car ownership and use between different countries (Table 11.6).

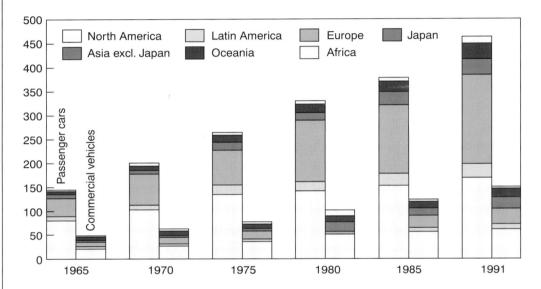

Fig. 11.6
Growth in passenger cars and commercial vehicles 1965-91[20].

Table 11.6
Car ownership in 1991 in some representative countries[21].

Country	Persons per car
USA	1.7
Italy	2.0
UK	2.4
Japan	3.3
Czechoslovakia	4.8
Malaysia	8.4
Republic of Korea	15.5
Ex-USSR	17.0
India	121.4
China	680.0

about £21 million on energy. Since 1980 the energy consumed per tonne of sugar has been reduced by 41 per cent through low grade heat recovery, cogeneration schemes and better control of heating and lighting[22]. Other potential decreases in carbon dioxide emissions can occur through the recycling of materials, the use of waste as an energy source and through switching to less carbon intensive fuels. Many studies in industrialized countries indicate that savings of 30 per cent or more could be made in the industrial sector at a net saving in overall economic terms[23].

In the developed world between one fifth and one quarter of energy use is for transport, and it is a rapidly growing proportion. Road transport accounts for the largest proportion of this (see box), over 80 per cent in industrialized countries; air transport is next at 13 per cent. In the United Kingdom, for instance, if the present trend in motor-car use continues, in about thirty years there will be twice as many cars doing twice as many miles – and the situation is very similar elsewhere. The proportional growth in developing countries could be even more rapid. The advantages conferred by the motor-car, the convenience, freedom and flexibility which it brings, mean that growth in its use is bound to continue. Increased prosperity also brings with it increased movement of freight.

There are two types of action which can be taken to curb the energy use of transport[24]. The first is to increase the efficiency of fuel use. We cannot expect the average car to compete with the vehicle which, in 1992, set a record by covering over 12,000 km on one gallon of petrol – a journey which serves to illustrate how inefficiently we use energy for transport! However, it is estimated that the average fuel consumption of the current fleet of motor cars could be halved through the use of existing technology – more efficient engines, lightweight construction and low-air-resistance design – while maintaining an adequate performance. The second action is to plan cities and other developments so as to lessen the need for transport and to make personalized transport less necessary – work, leisure and shopping should all be easily accessible by public transport, or by walking or cycling. Such planning needs also to be linked with a recognition of the importance of ensuring that public transport is reliable, convenient, affordable and safe.

Before leaving the question of energy efficiency, two other areas should be mentioned. Firstly, there are the questions of how far the efficiency of large power stations or other installations burning fossil fuels can be improved and also of whether their carbon dioxide emissions could be prevented from entering the atmosphere.

The efficiency of coal-fired power stations, for instance, has improved from about 32 per cent, a typical value of 20 years ago, to about 42 per cent for a pressurized fluidized bed combustion plant of today. Gas turbine technology has also improved providing efficiency improvements from about 36 per cent for an average plant to 45 per cent for the best modern plant. Such improvements are very significant in environmental terms and it is important that means be provided for the latest, most efficient technology to be available and attractive to rapidly industrializing countries such as China and India. Substantial further gains in overall efficiency can be realized by making sure that the large quantities of low-grade heat generated by power stations is not wasted but utilized, for instance in combined heat and power (CHP) schemes. For such cogeneration, the efficiencies attainable in the use of the energy from combustion of the fuel are typically around 80 per cent.

To prevent carbon dioxide from entering the atmosphere, it can either be removed from the flue gases or the fossil fuel feedstock could, in a gasification plant, be converted through the use of steam[25], to carbon dioxide and hydrogen. The carbon dioxide is then relatively easy to remove and the hydrogen can be used as a versatile fuel. The latter option will become more attractive when the technical problems of the large scale use of hydrogen in fuel cells to generate electricity have been overcome – we mention this again later in the chapter. Various options have been considered for the disposal of the very large amounts of carbon dioxide which result, for instance by pumping the carbon dioxide into spent oil or gas wells or into the deep ocean. In the most favourable circumstances (for instance when power stations are close to oil or gas fields), the cost of removal, although significant, might only be a small fraction of the total energy cost. For instance, in Norway where there is carbon tax of 15$US per tonne of carbon, a company is finding it economic to pump over one million tonnes per year of carbon dioxide removed from a natural gas stream into storage under the North Sea. In other circumstances and for deposition in the deep ocean the cost would be larger (perhaps up to 100 per cent on top of the energy cost). Further, if deposition in the deep ocean is seriously contemplated, the potential hazards associated with it need to be carefully assessed.

The other area needing brief consideration is that of the 10 per cent or so of world energy which is not commercial and which comes from traditional sources, largely from wood or other biomass. This is the main source of energy for half the world's population. Although all of these sources are renewable, it is still important that they are employed efficiently, and a great deal of room for increased efficiency exists. For instance, much cooking is still carried out on open fires where only about 5 per cent of the heat reaches the inside of the cooking pot. The introduction of a simple stove can increase this to 20 per cent or with a little elaboration to 50 per cent[26]. There is often considerable consumer resistance to the introduction of such stoves, but if they were widely used, the saving worldwide in fuelwood would be enormous. Other means of reducing fuelwood demand would be to encourage alternatives such as the use of fuel from crop wastes, of methane from sewage or other waste material or of solar cookers (mentioned again later on).

Most of the proposals described in this section fall into the category of 'no regrets' proposals – mentioned in Chapter 9. In other words, not only do they lead to substantial reductions in greenhouse gas emissions but they are good to do for other reasons – they lead to increased efficiency, cost savings or improvements in performance or comfort. It remains the case, however, that basic energy is generally so cheap, that without both encouragement and incentives, progress with the implementation of many of the proposals will be limited. Some of the policy instruments mentioned later could address this issue.

Table 11.7

Contributions to world energy supply (in millions of tonnes of oil equivalent) from renewable sources in 1990 and as assumed under the WEC scenario C in 2020[27].

	1990		2020	
	Mtoe	% of World energy	Mtoe	% of World energy
'Modern' biomass	121	1.4	561	5.0
Solar	12	0.1	355	3.1
Wind	1	0.0	215	1.9
Geothermal	12	0.1	91	0.8
'Small' hydro	18	0.2	69	0.6
Oceanic	0	0.0	54	0.5
Total (new renewable sources)	164	1.8	1,345	11.9
'Large' hydro	465	5.3	661	5.8
'Traditional' biomass	930	10.6	1,060	9.3
Total (all renewables)	1,559	17.7	3,093	27.0

Renewable energy

To put our energy use in context it is interesting to realize that the energy incident on the Earth from the sun amounts to about 180 thousand million million watts (or 18,000 Terawatts, 1 TW = 10^{12} W). This is about 15,000 times the world's average energy use of about 12 million million watts (12 TW). As much energy arrives at the Earth from the sun in 40 minutes as we use in a whole year. So, providing we can harness it satisfactorily and economically, there is plenty of renewable energy coming in from the sun to provide for all the demands human society can conceivably make.

There are many ways in which solar energy is converted into forms that we can use; it is interesting to look at the efficiencies of these conversions. If the solar energy is concentrated, by mirrors for instance, almost all of it can be made available as heat energy. Between 1 and 2 per cent of solar energy is converted through atmospheric circulation into wind energy, which although concentrated in windy places is still distributed through the whole atmosphere. About 20 per cent of solar energy is used in evaporating water from the Earth's surface which eventually falls as precipitation, giving the possibility of hydropower. Living material turns sunlight into energy through photosynthesis with an efficiency of around 1 per cent for the best crops. Finally, photovoltaic (PV) cells convert sunlight into electricity with an efficiency which for the best modern cells can be over 20 per cent.

Around the year 1900, very early on in the production of commercial electricity, water power was an obvious source and from the beginning made an important contribution. Hydroelectric schemes now supply about 6 per cent of the world's commercial energy. Other renewable sources of commercial energy, however, have been dependent on recent technology for their implementation. In 1990, only about 2 per cent of the world's commercial energy came from renewable sources other than large hydro[28] (these are often collectively known as 'new renewables'). Of this 2 per cent (Table 11.7), about three-quarters was from 'modern' biomass (called 'modern' when it contributes to commercial energy to distinguish it from traditional biomass), the other 0.5 per cent being shared between solar, wind energy, geothermal and small hydro sources.

Returning to commercial energy generation, in order to put renewable sources into context, it is useful to inspect the detailed projection of the WEC (Table 11.7) for the contributions from different 'new renewable' sources which make up the 12 per cent of total energy supply in the year 2020 assumed for the WEC scenario C. The main growth expected is in energy from 'modern' biomass and from solar and wind energy sources.

In the following paragraphs, the main renewable sources are described in turn and their possibilities for growth considered[29]. Most of them are employed for the production of electricity through mechanical means (for hydro and wind power), through heat engines (for biomass and solar thermal) and through direct conversion from sunlight (solar PV). In the case of biomass, liquid or gaseous fuels can also be produced.

Hydro-power

Hydro-power, the oldest form of renewable energy, is well established and is competitive economically with electricity generated by other means. Some hydroelectric schemes are extremely large. The world's largest, the Three Gorges project on the Yangtze river in China, when completed will generate about 20,000 MW of electricity. Two other large schemes, each of over 10,000 MW capacity, are in South America at Guri in Venezuela and at Itaipu on the borders of Brazil and Paraguay. It is estimated[30] that there is potential for further exploitation of hydroelectric capacity to three or four times what has currently been developed, much of this undeveloped potential being in the former Soviet Union and in developing countries. Large schemes, however, can have significant social impact (such as the movement of population from the reservoir site), environmental consequences (for example, loss of land, of species and of sedimentation to the lower reaches of the river), and problems of their own such as silting up, which have to be thoroughly addressed before they can be undertaken.

But hydroelectric schemes do not have to be large; Table 11.7 distinguishes between large and small hydroelectric sources. Many units exist generating a few kilowatts only which may supply one farm or a small village. The attractiveness of small schemes is that they provide a locally based supply at modest cost. For instance, in China there are a large number of small hydro plants of less than 10 MW which aggregated in 1990 to some 4,000 MW in total capacity; a further 2,000 MW of such schemes are currently under development in that country. Many more possibilities exist for the exploitation of the potential of small rivers and streams in many parts of the world. About 1,000 MW capacity per year is currently being installed globally in small hydro schemes.

Given appropriate incentives by governments and the availability of finance it is estimated that this rate of installation could double[31].

An important facility provided by some hydro schemes is that of pumped storage. Using surplus electricity available in off-peak hours, water can be pumped from a lower reservoir to a higher one. Then, at other times, by reversing the process, electricity can be generated to meet periods of peak demand. The efficiency of conversion can be as high as 80 per cent and the response time a few seconds, so reducing the need to keep other generating capacity in reserve. In 1990 about 75,000 MW of pumped storage capacity was available worldwide with a further 25,000 MW under construction[32].

Biomass as fuel

Second in current importance as a renewable energy source is the use of biomass as a fuel. The word biomass in this context covers domestic, industrial and agricultural dry waste material, wet waste material and crops, all of which can be used as fuel for heating and to power electricity generators; some are also appropriate to use for the manufacture of liquid or gaseous fuels.

There is considerable public awareness of the vast amount of waste produced in modern society. The United Kingdom, for example, produces each year somewhat over 30 million tonnes of domestic solid waste, or about half a tonne for every citizen; this is a typical value for a country in the developed world. Even with major programmes for recycling some of it, large quantities would still remain. If it were all incinerated for power generation (modern technology enables this to be done with negligible air pollution) about 1.7 GW could be generated, about 5 per cent of the UK's electricity requirement[33]. Uppsala in Sweden is an example of a city with a comprehensive district heating system, for which, before 1980, over 90 per cent of the energy was provided from oil. A decision was then made to move to renewable energy and by 1993, energy from waste incineration and from other biomass fuel sources provided nearly 80 per cent of what is required for the city's heating.

But what about the greenhouse gas generation from waste incineration? Carbon dioxide is of course produced from it, which contributes to the greenhouse effect. However, the alternative method of disposal is landfill (most of the waste in the UK currently is disposed of that way). Decay of the waste over time produces carbon dioxide and methane in roughly equal quantities. Some of the methane can be collected and used as a fuel for power generation. However, only a fraction of it can be captured; the rest leaks away. Because methane is a much more effective greenhouse gas, molecule for molecule, than carbon dioxide, the leaked methane makes a substantial contribution to the greenhouse effect. Detailed calculations show that if all United Kingdom domestic waste were incinerated for power generation rather than landfilled, the net saving per year in greenhouse gas emissions would be equivalent to about 10 million tonnes of carbon as carbon dioxide[34]. Since this is about 5 per cent of the total United Kingdom greenhouse gas emissions, we can infer that power generation from waste could be a significant contribution to the reduction in overall emissions. These figures for electricity generation and saving in greenhouse gases could be about doubled if in addition to the domestic waste the potential for power generation from industrial and agricultural waste was also taken into account. An idea of the cost of electricity provided from these sources is shown in Fig. 11.7.

Other wastes resulting from human or agricultural activity are wet wastes such as sewage sludge and farm slurries and manures. Bacterial fermentation in the absence of oxygen (anaerobic digestion) of these wastes produces biogas which is mostly methane and which can be used as a fuel to produce energy. Although not a large potential energy source (Fig. 11.7), there is room for its contribution to increase.

The third potential biomass contribution is from the use of crops as a fuel. Here the potential is large. It is a genuinely renewable resource in that the carbon dioxide which is emitted when the biomass is burnt is turned back again into carbon, through the process of photosynthesis, in the renewed biomass when it is grown again. Many different crops can be employed as biomass for energy production. In Brazil, for instance, since the 1970s large plantations of sugar cane have produced alcohol for use as a fuel mainly in transport, generating, incidentally, much less local pollution than petrol or diesel fuel from fossil sources. A lot of potential has been recognized for the sugar cane industry to produce both sugar and energy together with other byproducts as well (see box). Biomass from wood plantations on agricultural land no longer needed for food crops features as an important future source in Sweden's energy plans[36]; the most efficient use of the biomass is first to turn it into biogas and then burn it in a gas turbine to produce electricity. For the UK, trials indicate that the most promising option is willow and poplar grown in coppices[37].

Because of the low efficiency of conversion of solar energy to biomass, the amount of land required for significant energy production by this means is large – and it is important that land is not taken over which is required for food production. However, there is in principle no shortage of land for this purpose. Plenty of suitable crops are available which could be grown on land only

Fig. 11.7
Estimated contribution to electricity generation in the UK from renewable energy sources in the year 2025 as a function of the electricity price with an assumption of an 8% discount rate. The assessment for electricity generated from crops is considered a preliminary one[35].

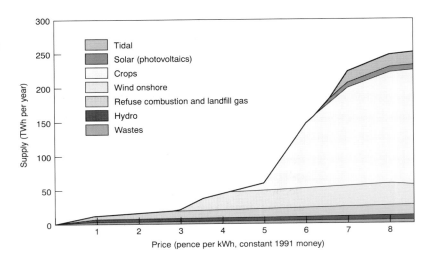

Sugar cane as biomass

A sugar cane factory produces many different byproducts which can be efficiently employed as sources of energy – either for biofuels or for electricity production.

Sugar cane production yields two kinds of biomass fuel suitable for gasification (Fig. 11.8), known as bagasse and barbojo. Bagasse is the residue from crushing the cane and is thus available during the milling season; bar-bojo consists of the tops and leaves of the cane plant that could be stored for use after the milling season. It has been estimated that using these sugar cane sources, within thirty years or so, the 80 sugar cane producing countries in the developing world could generate two-thirds of their current electricity needs at a price competitive with fossil fuel energy sources[38].

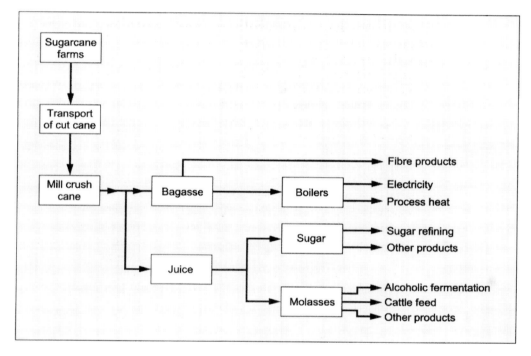

Fig. 11.8
The processes involved in the sugar cane industry[39].

marginally useful for agriculture. In many developing countries biomass plantations can provide suitable fuel for local electricity generation more competitively than other means of generation.

In industrialized countries the main drawback is the cost of energy production from biomass; the cost of electricity generated this way is currently typically about twice that generated from fossil fuel sources, and the cost of alcohol from biomass as a fuel for transport is about twice the basic cost of petrol or diesel. However, with appropriate development aimed at selecting the most suitable crops and generation technologies, these costs will be reduced. Further, in some countries the expenditure which would otherwise be incurred in the setting aside of land from agricultural use for food production could be set against that extra cost.

Wind energy

Energy from the wind is not new. Two hundred years ago windmills were a common feature of the European landscape; for example, in 1800 there were over 10,000 working windmills in Britain. During the past few years they have again become familiar on the skyline especially in countries in western Europe (for instance, Denmark, Great Britain and Spain) and in western North America. Slim, tall, sleek objects silhouetted against the sky, they do not have the rustic elegance of the old windmills, but they are much more efficient. A typical large wind energy generator will have a two- or three-bladed propeller about 50 m in diameter and a rate of power generation in a wind speed of 12 m/sec (43 km/hr, 27 mph or Beaufort Force 6), of about 700 kilowatts. On a site with an average wind speed of about 7.5 m/sec (an average value for exposed places in many western regions of Europe) it will generate an average power of about 250 kW. The generators are often sited close to each other in wind farms which may include several dozen such devices.

From the point of view of the electricity generating companies the difficulty with the generation of electrical power from wind is that it is intermittent. There are substantial periods with no generation at all. The generating companies can cope with this in the context of a national electricity grid which pools electrical power from different sources providing that the proportion from wind sources is not too large.

The power generated from the wind depends on the cube of the wind speed: a wind speed of 12.5 m/sec is twice as effective as one of 10 m/sec. It makes sense therefore to mount wind farms on the windiest sites available. For instance, some of the strongest winds in Europe are found near the west coasts of Great Britain where it has been estimated that wind energy could within reasonable cost (Fig. 11.7) contribute up to 10 per cent of the UK's electricity supply[40]. Such a provision would require more than 20,000 700 kW generators; these would need to be carefully sited if visual amenity were not to be lost. It would also require that the electricity grid be reinforced to receive power generated in remote but windy regions of western Scotland or from wind farms mounted off-shore. In the longer term this provision could perhaps be doubled, especially if an effective means for energy storage (for instance using hydrogen; more of that possibility later in the chapter) were developed. Many other countries with windy coast lines or mountain regions are also beginning to make substantial use of wind energy. In Denmark, by 1996 800 MW capacity had been installed contributing 4.5 per cent to the country's total electricity requirement (see box on p. 222). Worldwide the total capacity of wind generators increased by about 20 per cent in 1996 to about 6,000 MW. Rapid growth in wind power capacity could be achieved in many countries. For instance, it has been estimated that India could be generating up to 10,000 MW of electrical power (about a quarter of current needs) from wind by 2030[41].

Wind power on Fair Isle

A good example[42] of a site where wind power has been put to good effect is Fair Isle, an isolated island in the North Sea north of the Scottish mainland. Until recently, the population of 70 depended on coal and oil for heat, petrol for vehicles and diesel for electricity generation. A 50 kW wind generator was installed in 1982 to generate electricity from the persistent strong winds of average speed over 8 m/sec (29 km/h or 18 mph). The electricity is available for a wide variety of purposes; at a relatively high price for lighting and electronic devices and at a lower price controlled amounts are available (wind permitting) for comfort heat and water heating. At the frequent periods of excessive wind further heat is available for heating glasshouses and a small swimming pool. Electronic control coupled with rapid switching enables loads to be matched to the available supply. An electric vehicle has been charged from the system to illustrate a further use for the energy.

With the installation of the wind generator, which now supplies over 90 per cent of the island's electricity, electricity consumption has risen about fourfold and the average electricity costs have fallen from 13p/kWh to 4p/kWh. A second wind turbine of 100 kW capacity was installed in 1996/7 to meet increasing demand and to improve wind capture.

Wind energy is also particularly suitable for the generation of electricity at isolated sites to which the transmission costs of electricity from other sources would be unacceptable. Because of the wind's intermittency, some storage of electricity or some back-up means of generation has to be provided as well. The installation on Fair Isle (see box) is a good example of an efficient and versatile system. Small wind turbines also provide an ideal means for charging batteries in isolated locations; for instance, about 83,000 are in use by Mongolian herdsmen. Wind energy is often also an ideal source for water pumps.

With the large growth during the last few years in the installations of wind generators, economies of scale have brought down the cost of the electricity generated so that it is approaching the cost of electricity generated from fossil fuels (Fig. 11.7).

Energy from the sun

The simplest way of making use of energy from the sun is to turn it into heat. A black surface directly facing full sunlight can absorb about 1 kilowatt for each square metre of surface. In countries with a high incidence of sunshine it is an effective and cheap means of providing domestic hot water which is extensively employed in countries such as Australia, Israel, Japan and the southern states of the United States of America (see box). In tropical countries, a solar cooking stove can provide an efficient alternative to stoves burning

Solar water heating

The essential components of a solar water heater (Fig. 11.9) are a set of tubes in which the water flows embedded in a black plate insulated from behind and covered with a glass plate on the side facing the sun. A storage tank for the hot water is also required. A more efficient (though more expensive) design is to surround the black tubes with a vacuum to provide more complete insulation.

Fig. 11.9
Design of solar water heater: a solar collector connected to a storage tank through a circulating pump. Alternatively, if the storage is above the collector, the hot water will collect through gravity flow[43].

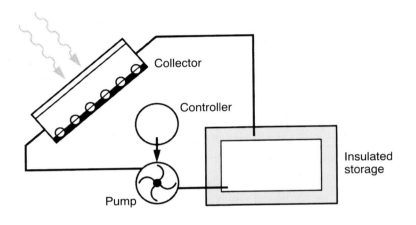

wood and other traditional fuels. Thermal energy from the sun can also be employed effectively in buildings (it is called passive solar design), in order to provide a modest boost towards heating the building in winter and, more importantly, to provide for a greater degree of comfort and a more pleasant environment (see box opposite).

Solar heat can also be employed to provide heating to produce steam for the generation of electricity. To produce significant quantities of steam, the solar energy has to be concentrated by using mirrors. One arrangement employs trough-shaped mirrors aligned east-west which focus the sun on to an insulated black absorbing tube running the length of the mirror. A number of such installations have been built, particularly in the United States, where solar thermal installations provide over 350 MW of commercial electricity. The high capital cost of such installations, however, assuming a reasonable pay-back period, translates into an electricity cost which, at the moment, is at least three times that from most conventional sources. Generating plants which incorporate integrated solar and fossil fuel heat sources in combined cycle operation are currently under development which promise significantly lower costs[44].

Sunlight can be converted directly into electricity by means of photovoltaic (PV) solar cells (see box on p. 212). They appear in a variety of ways in everyday life; as power sources, for instance for small calculators or watches. Spacecraft are covered with 'solar panels' or possess 'solar arrays' covered with PV cells providing electrical power for the spacecraft. Their efficiency for conversion of

Solar energy in building design

All buildings benefit from unplanned gains of solar energy through windows and, to a lesser extent, through the warming of walls and roofs. This is called 'passive solar gain'; for a typical house in the United Kingdom it will contribute about 15 per cent of the annual space heating requirements. With 'passive solar design' this can relatively easily and inexpensively be increased to around 30 per cent, while increasing the overall degree of comfort and amenity. The main features of such design are to place, so far as is possible, the principal living rooms with their large windows on the south side of the house, with the cooler areas such as corridors, stairs, cupboards and garages with the minimum of window area arranged to provide a buffer on the north side. Conservatories can also be strategically placed to trap some solar heat in the winter.

The wall of a building can be designed specifically to act as a passive solar collector, in which case it is known as 'solar wall' (Fig. 11.10)[45]. Its construction enables sunlight, after passing through an insulating layer, to heat the surface of a wall of heavy building blocks which retain the heat and slowly conduct it into the building. The insulating layer, although allowing sunlight to pass through, prevents thermal radiation from passing out. A retractable reflective blind can be placed in front of the insulation at night or during the summer when heating of the building is not required. A set of student residences for 376 students at Strathclyde University in Glasgow in southwest Scotland has been built with a 'solar wall' on its south-facing side. Even under the comparatively unfavourable conditions during winter in Glasgow (the average duration of bright sunshine in January is only just over one hour per day) there is a significant net gain of heat through the wall to the building.

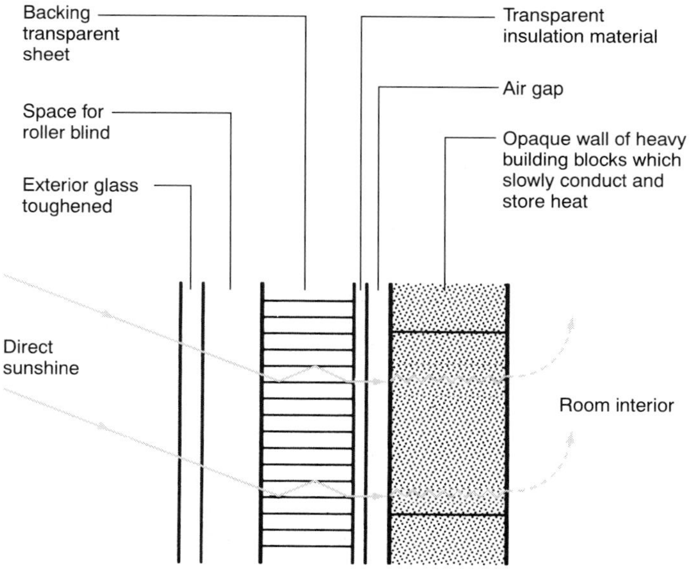

Fig. 11.10
Construction of a 'solar wall'. The insulation material is about 100 mm thick and consists of open honeycomb channels of transparent polycarbonate material.

Backing transparent sheet

Space for roller blind

Exterior glass toughened

Direct sunshine

Transparent insulation material

Air gap

Opaque wall of heavy building blocks which slowly conduct and store heat

Room interior

solar energy into electrical energy is now generally between just under 10 per cent and 20 per cent. A panel of cells of area one square metre facing full sunlight will therefore deliver between 100 and 200 watts of electrical power.

The photovoltaic solar cell

The silicon photovoltaic (PV) solar cell consists of a thin slice of silicon into which appropriate impurities have been introduced to create what is known as a p–n junction. The most efficient cells are sophisticated constructions using crystalline silicon as the basic material; they possess efficiencies for the conversion of solar energy into electricity typically of 15–20 per cent; experimental cells have been produced with efficiencies well over 20 per cent[46].

Single crystal silicon is less convenient for mass production than amorphous silicon (for which the conversion efficiency is around 10 per cent) which can be deposited in a continuous process on to thin films[47]. Other alloys (such as cadmium telluride and copper indium selenide) with similar photovoltaic properties can also be deposited in this way and, because they have higher efficiencies than amorphous silicon, may become important in the future[48].

Cost is of critical importance if PV solar cells are going to make a significant contribution to energy supply. This has been coming down rapidly. More efficient methods and larger-scale production are bringing the cost of solar electricity down to levels where it can compete with other sources. Projections up to the year 2020 of the likely cost of electricity generated from PV sources are compared with that from conventional sources in Fig. 11.11.

Fig. 11.11
The falling cost of electricity produced from PV solar cells as estimated by the World Energy Council[49]. Projections of the range of costs of PV electricity production is compared with the range of cost of electricity from conventional fossil fuel sources.

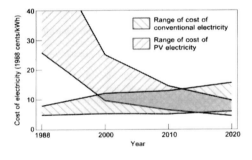

Installed on rooftops in cities they provide a way for city dwellers to contribute renewably to their energy needs (in Japan about 4,000 roof systems towards a target of 70,000 had been installed by the end of 1996). The cost of energy from solar cells has reduced dramatically over the past twenty years (see box); so much so that they can now be employed for a wide range of applications and can also begin to contribute to the large-scale generation of electricity.

Small PV installations are also suitable, especially for developing countries, to provide local sources of electricity in rural areas. About a third of the world's population have no access to electricity from a central source. Their predominant need is for small amounts of power for lighting, for radio and television, for refrigerators (for example, for vaccines at a health clinic) and for pumping water. The cost of PV installations for these purposes is now competitive with other means of generation (such as diesel units) and by 1992 around 100,000 DLKs (Domestic Lighting and broadcast reception charging Kits – Fig. 11.12) had been installed worldwide[50]. Substantial use of solar PV is being made by African countries, for instance about 1MW had been installed in Kenya by 1993, mostly in small installations. Larger installations are required for public buildings, although they need not be that much larger. Many small hospitals can benefit from an electrical power source as small as 1 to 2 kW. For instance, by 1995, 70 small hospitals in Sri Lanka, through assistance from the Australian government, had installed 1.3 kW solar arrays, backed up by 2,200 amp-hour batteries, to provide for lighting, refrigeration for vaccines, autoclave sterilization, pumping for hot water (produced through a solar thermal system) and radio.

In 1996 the world capacity for production of solar cells was equivalent to a peak electrical power of about 60 MW. To meet the estimates under the WEC scenario C of their projected contribution to world energy supply of at least 100 GW[51] by the year 2020 (Table 11.7), production will need to increase by over a factor of a thousand. That means an average increase of over 30 per cent per annum (compare the increase of about 15 per cent for 1995/6). In the short term, increased development of local installations is likely to have priority; after the turn of the century, with the expectation of a significant cost reduction (Fig. 11.11), penetration into large-scale electricity generation will become more possible.

Other renewable energies

We have so far covered the renewable energy sources for which there is potential for growth on a scale which can make a significant contribution to overall

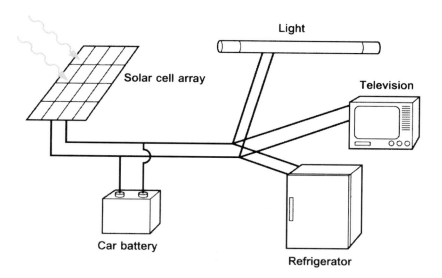

Fig. 11.12
A simple solar power installation for a small home is being marketed in Sri Lanka for less than $(US)100[52]. An array of 36 solar cells, each 10 cm in diameter, provides 40 watts of peak power. This is sufficient to charge a car battery which can power up to three 9 watt fluorescent lights and three hours of radio and one hour of television each day. With more restricted use of these devices or with a larger solar array, a small refrigerator can be added to the system.

world energy demand. We should also mention briefly other renewable energy technologies which contribute to global energy production and which are of particular importance in certain regions, namely geothermal energy from deep in the ground and energies from the tides, waves or thermal gradients in the ocean.

The presence of geothermal energy from deep down in the Earth's crust makes itself apparent in volcanic eruptions and less dramatically in geysers and hot springs. The energy available in favourable locations may be employed directly for heating purposes or for generating electrical power. Although very important in particular places, for instance in Iceland, it is currently only a small contributor (between 0.1 and 0.2 per cent) to total world energy; its contribution could rise to the order of 1 per cent during the next few decades (Table 11.7).

Large amounts of energy are in principle available in the movements and the temperature gradients of the ocean; but they are in general not easy to exploit. Tidal energy is the only one currently contributing significantly to commercial energy production. The largest tidal energy installation is at La Rance in France; it has a capacity of 240 MW. Several estuaries in the world have been extensively studied as potential sites for tidal energy installations. The Severn Estuary in the United Kingdom, for instance, has the potential to generate a peak power of over 8,000 MW or about 6 per cent of the total UK electricity demand. Although the long term cost of the electricity generated from the largest schemes could be competitive, the main deterrents to such schemes are the high capital up-front cost and the significant environmental impacts that can be associated with them.

The financing of renewable energy

Renewable energy on the scale envisaged by WEC scenario C will only be realized if it is seen to be competitive in cost with energy from other sources. In due course, as easily recoverable oil and gas reserves begin to run out, those fuels will become more expensive enabling renewable sources to compete more easily[53]. The earliest that could occur is some decades away and, since estimates of recoverable fossil fuel reserves have always tended to be too low, it may be well into the second half of next century before any substantial limitation in oil and gas resources occur. Before then other means to address the competitiveness of renewable energies and the investment they require will have to be found.

Considering electricity and the cost of various sources of its supply, in Fig. 11.13 comparison is made between these costs (for the USA) and the carbon dioxide emissions from the different sources. Under some circumstances renewable energy sources are already competitive in cost, for instance in providing local sources of energy where the cost of transporting electricity or other fuel would be significant; some examples of this (such as Fair Isle in Scotland) have been given. However, if renewables are to begin to displace fossil fuels to the extent required by scenario C, there will need to be financial incentives to bring about the change.

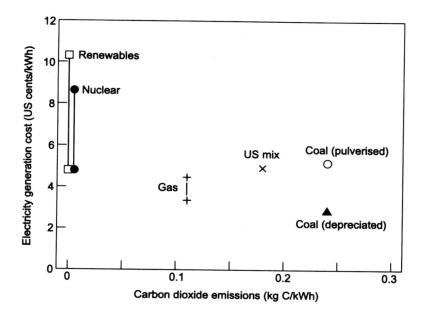

Fig. 11.13

Illustrating the cost-effectiveness of electricity supply options: the cost of electricity generation in the United States from various sources plotted against their generation of carbon dioxide emissions (expressed in terms of kg of carbon as carbon dioxide). Also plotted is the figure representing the average for generation in the United States in 1990 (the US mix). Distinction is made between coal generating plants which have fully depreciated and new pulverized coal plants. A range of values is given for nuclear energy and for renewable energy sources[54].

At the basis of such incentives would be the principle that the polluter should pay by the allocation of an environmental cost to carbon dioxide emissions. Supposing, for instance, that through taxes or levies an additional cost of between 50 and 100$US per tonne of carbon (figures mentioned in the context of environmental costs towards the end of Chapter 9) were to be associated with carbon dioxide emissions, between 0.5 and 2.5 cents per kWh would be added to the price of electricity from fossil fuel sources (Fig. 11.13) – which could begin to bring some renewables (for instance, biomass and wind energy) into competition with them. It is interesting to note that in many countries substantial subsidies are attached to energy – worldwide they amount on average to the equivalent of about 40$US per tonne of carbon[55]. A start with incentives would therefore be made if subsidies were removed from energy generated from fossil fuel sources.

Another means of favouring renewable energy over against energy from fossil fuel sources could be through tradeable permits in carbon dioxide emissions. These control the total amount of carbon dioxide which a country or region may emit while providing the means for industries to trade permits for their allowable emissions within the overall total.

Electricity, however, only accounts for about one third of the world's primary energy use. What about the possibility of renewable energy replacing solid and liquid fuels used for heating, for industry and for transport? It has already been mentioned that currently, liquid fuels such as ethanol derived from biomass are about twice as expensive as those derived from oil. Although there is an expectation that the processing of biomass will become more efficient[56] – the rapid development of technologies in bioengineering will help – it is unlikely that in the short term the substitution of biomass-derived fuels will occur on a large scale without the application of appropriate financial incentives. Arrangements of a similar kind to those we have already mentioned for electricity production would need to be put into place.

Looking at the investment required, annual investment in the world's energy industry is already large; it currently runs at between 3 and 4 per cent of GWP. The WEC estimates that cumulative world investment in energy supply will continue at least at the same level and up to the year 2020 will approach 30 million million US dollars at 1992 prices. Large investment will be required to meet the growth rates in the implementation of renewable energy necessary to achieve the projections, for instance, of WEC scenario C – a growth of 30 per cent per annum in solar cell production, for instance, was mentioned earlier. The WEC estimate that investment of at least 2.4 million million US dollars in new renewable energy sources would be required up to 2020. Although this latter sum is less than 10 per cent of the world's total investment in energy, appropriate encouragement and incentives will be required if it is to be realized. Some of the policy instruments which are mentioned below would need to be applied.

Nuclear energy

An energy source mentioned rather little so far is nuclear energy. It is not strictly a renewable source, but it has considerable attractiveness from the point of view of sustainable development because it does not produce greenhouse gas emissions (apart from a small amount which is used in making the materials employed in the construction of nuclear power stations) and because the rate at which it uses up resources of radioactive material is small compared to the total resource available. It is only efficiently generated in large units, so is suitable for supplying power to national grids or to large urban connurbations, but not for small, more localized supplies. The cost of nuclear energy compared with energy from fossil fuel sources (Fig. 11.13) is often a subject of debate; exactly where it falls in relation to the others depends on the return expected on the up-front capital cost and on the cost of decommissioning of spent power stations, which represent a large element of the total. A further advantage of nuclear energy installations is that the technology is known; they can be built now and therefore contribute to the reduction of carbon dioxide emissions in the short term.

The continued importance of nuclear energy is recognized in the WEC energy scenarios, which all assume growth in this energy source next century. How much growth will be realized will depend to a large degree on how well the nuclear industry is able to satisfy the general public of the safety of its operations; in particular that the risk of accidents from new installations is negligible, that nuclear waste can be safely disposed of, that the distribution of dangerous nuclear material can be effectively controlled and that it can be prevented from getting into the wrong hands.

A further potential nuclear energy source depends on fusion rather than fission. When, at extremely high temperatures, the nuclei of hydrogen (or one of its isotopes) are fused to form helium, a large amount of energy is released. This is the energy source which powers the sun; if a suitable method for harnessing and containing the process on the Earth could be devised, virtually limitless supplies of energy could be provided. A great deal of research is being put into fusion technology, but the economic generation of power by this means still appears to be many decades away.

Technology for the longer term

This chapter has concentrated mostly on what can be achieved with available and proven technology during the next few decades. It is also interesting to speculate about the more distant future and what relatively new technologies may become dominant later next century. In doing so, of course, we are almost certainly going to paint a more conservative picture than will actually occur. Imagine how well we would have done if asked in 1890 to speculate about technology change by the 1990s! Technology will certainly surprise us with possibilities not thought of at the moment. But that need not deter us from being speculative!

Of the renewable energy sources mentioned above, the direct conversion of sunlight into electricity by photovoltaic (PV) cells is in many ways the most attractive. There are many regions of the world where sunshine is plentiful and where suitable land not useful for other purposes would be readily available. It is a very clean, non-polluting technology, easily adaptable to mass production. The cost of PV electricity has been coming down rapidly – a trend which is likely to continue especially with an increased scale of production.

But there are problems. Power from PV installations is only available during the day and such installations will usually be remote from places where the electricity is to be used. Means of energy storage and transport are therefore required. An important possibility here is for hydrogen gas to be used as a storage medium. It can be generated directly from the PV electricity by the electrolysis of water (Fig. 11.14). This can be a very efficient process; over 90 per cent of the electrical energy can be stored in the hydrogen. The hydrogen can then be transported by pipeline or by bulk transport. Again, hydrogen is an extremely non-polluting energy source, which can easily be applied to most of the uses for which energy is required.

An attractive means to turn the energy in the hydrogen into electricity is through the use of a hydrogen-oxygen fuel cell (see box) in which the electrolytic process of generating hydrogen and oxygen from water is reversed – the energy released by recombination of the hydrogen and oxygen is turned back into electrical energy. Fuel cells can have high efficiency of 50–80 per cent; they are pollution free and so would be very suitable for transport vehicles. Their disadvantage is that as yet they tend to be large, heavy and expensive

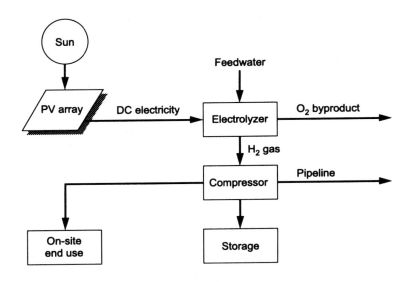

Fig. 11.14
A solar photovoltaic (PV) electrolytic hydrogen system[57].

Fuel cell technology

A fuel cell converts the chemical energy of a fuel directly into electricity without first burning it to produce heat. It is similar to a battery in its construction. Two electrodes (Fig. 11.15) are separated by an electrolyte which transmits ions but not electrons. A fuel cell has a theoretical efficiency of 100 per cent. Fuel cells have been constructed with efficiencies in the range of 40–80 per cent.

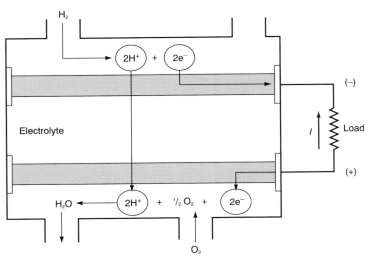

Fig 11.15
Schematic of a hydrogen-oxygen fuel cell[58]. Hydrogen is suppled to the porous anode (negative electrode) where it dissociates into hydrogen ions (H⁺) and electrons. The H⁺ ions migrate through the electrolyte (typically an acid) to the cathode (positive electrode) where they combine with electrons (supplied through the external electrical circuit) and oxygen to from water.

Estimates are that it will be 10–20 years before they have been developed sufficiently to be attractive for widespread use, but there seems general agreement that its potential advantages are such that the fuel cell will become an important technology for the longer term.

All the technology necessary for a solar-hydrogen energy economy is available now, although the cost of energy supplied this way would at the moment be several times that from fossil fuel sources[59]. As the technology for its further development progresses and as larger-scale production becomes possible, the cost will undoubtedly reduce substantially. If its attractiveness from an environmental point of view were recognized as a dominant reason for its rapid development, the solar-hydrogen economy could take off more rapidly than most energy analysts are currently predicting.

To assist in understanding the long-term potential of the different technologies and means of energy supply we have presented in previous sections, the IPCC 1995 Report put together alternative constructions (called Low-Emissions Supply System (LESS) constructions – see box p.220) of future energy supply options. All of them allow world energy use to grow during the century and allow also for deep reductions in carbon dioxide emissions from fossil fuel combustion similar to the projections in the WEC-C scenario, details of which were given earlier in the chapter. An important finding of the LESS construction exercise is that alternative strategies exist for the development and employment of technologies next century which would enable large reductions in fossil fuel carbon dioxide emissions to be achieved.

Policy instruments

Action in the energy sector on the scale required to mitigate the effects of climate change through reduction in the emissions of greenhouse gases will require significant policy initiatives by governments in cooperation with industry. Substantial investment is needed in research and development associated with renewable energy sources and efficient energy use; barriers to the diffusion and transfer of technology need to be reduced, financial resources need to be mobilized, supporting capacity needs to be built in developing countries and approaches need to be developed which will assist in the implementation of behavioural changes and technological opportunities in all regions of the globe.

The optimum mix of policies will vary from region to region and country to country. Some countries, for instance Denmark (see box p. 222), have had a strategy for sustainable energy in place for some time. An encouraging feature of international discussions is the willingness of developed and developing countries to work together on policy options, for instance in schemes of 'joint implementation' in which two countries consider the optimum ways of reducing their combined emissions of greenhouse gases.

The policy options available include[60]:

- putting in place appropriate institutional and structural frameworks;
- energy pricing strategies (carbon or energy taxes and reduced energy subsidies);

The Low-Emissions Supply System (LESS) Projections

In the IPCC 1995 Report are presented[61] alternative constructions of future energy supply options all of which would achieve deep reductions in carbon dioxide emissions by the end of next century. Emphasis in the constructions is on the long term – 2025 to 2100. During this period the global commercial energy system will have been replaced two or three times – providing many opportunities to change the system through the use of new technologies.

Details of the constructions are given in Figs. 11.16 and 11.17. All of them allow world energy use to grow during the century – by about a factor of two – and allow also for reduction in carbon dioxide emissions from fossil fuel combustion from about 6 Gt of carbon per annum in 1990 to about 4 Gt in 2050 and to about 2 Gt in 2100 – similar to the projections in the WEC-C scenario (details of which were given earlier in the chapter), similar too to the emission profiles which would lead to stabilization of atmospheric carbon dioxide concentration at around 450 ppmv (see Chapter 10).

The LESS constructions assume a world population of 10.5 billion by 2100, average world economic growth next century of about 3.3 per cent per annum and average energy intensity reduction of about 2.5 per cent per annum – similar to those of WEC scenario C (see box earlier in the chapter). The alternative constructions emphasize different basic sources of energy – biomass, nuclear, natural gas and coal. A fifth alternative allows for a much larger energy demand. In all of them, however, biomass plays a major role either as a source for electricity production or a feedstock for bio-fuels. Hydrogen is an important energy carrier in the longer term. When there is continued use of fossil fuels, it is assumed that they are employed at the greatest possible efficiency; it is also assumed especially in the fourth and fifth alternatives that large quantities of carbon dioxide are sequestered in oil or natural gas fields or elsewhere.

Fig 11.16.
Global primary energy use (in exajoules – see caption to Fig. 11.1 for definition) for the LESS constructions showing various options for future energy supply.

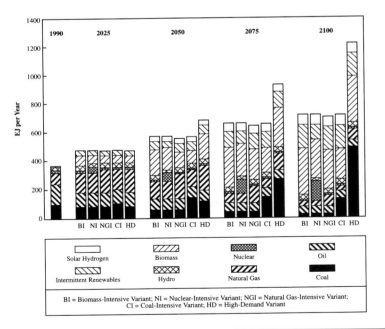

The Low-Emissions Supply System (LESS) Projections *continued*

Fig 11.17.
Annual carbon dioxide emissions from fossil fuels for the different LESS constructions shown in Fig 11.16, with comparison to the IS 92 scenarios[62].

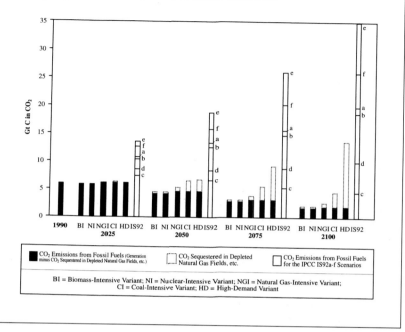

- CO$_2$ Emissions from Fossil Fuels (Generation minus CO$_2$ Sequestered in Depleted Natural Gas Fields, etc.)
- CO$_2$ Sequestered in Depleted Natural Gas Fields, etc.
- CO$_2$ Emissions from Fossil Fuels for the IPCC IS92a-f Scenarios

BI = Biomass-Intensive Variant; NI = Nuclear-Intensive Variant; NGI = Natural Gas-Intensive Variant; CI = Coal-Intensive Variant; HD = High-Demand Variant

---

- reducing or removing other subsidies (eg agricultural and transport subsidies) that tend to increase greenhouse gas emissions;
- tradeable emissions permits;
- voluntary programmes and negotiated agreements with industry;
- utility demand-side management programmes;
- regulatory programmes, including minimum energy efficiency standards (eg for appliances and fuel economy);
- stimulating research and development to make new technologies available;
- market pull and demonstration programmes that stimulate the development and application of advanced technologies;
- renewable energy incentives during market build-up;
- incentives such as provisions for accelerated depreciation or reduced costs for consumers;
- education and training; information and advisory measures;
- options that also support other economic and environmental goals.

Summary

This chapter has outlined the ways in which energy for human life and industry is currently provided. Growth in conventional energy sources at the rate required to meet future energy needs will generate much increased emissions of greenhouse gases. The WEC have taken a lead by proposing a future energy scenario (scenario C) driven strongly by environmental considerations

Energy strategy in Denmark

Denmark is a country which has developed a sustainable energy strategy which includes a target for the reduction of carbon dioxide emissions from fossil fuel sources to 50 per cent of 1988 levels by 2030. Key components of the strategy are:

- an emphasis on energy efficiency, through tough building standards regulating building insulation (eg the inclusion of 250 mm of insulation in wall construction) and design (eg the inclusion of passive solar design) and through district heating sytems which utilize waste heat from power stations, waste incineration, sewage and landfill gas combustion etc. Denmark has one of the world's lowest values of energy intensity (ratio of energy demand to GDP) – 25 per cent less than Japan and over 60 per cent less than the USA.
- a strategy for renewable energy, in particular from wind, biomass and the use of rural and urban wastes. A key component of the strategy has been the stimulation of markets for assisted but competitive manufacture and the use of taxes on gaseous emissions of carbon, nitrogen and sulphur compounds from fossil fuels.
- growth in the wind power industry of over 10 per cent per annum since 1975 leading to the installation by 1996 of 800 MW of wind turbine capacity, providing 4.5 per cent of the country's total electricity. Annual exports of wind turbines were worth about 500 million $(US) in 1995 and provided employment for over 5,000 people.
- the generation of a significant proportion of primary energy for electricity and heat from biomass and waste incineration, usually combined with district heating. By 1996, 8 per cent of the total requirement of this primary energy was provided from renewable sources. Substantial growth in energy from these sources must continue into the future if the target for carbon dioxide emissions is going to be met.

which would lead to the stabilization of carbon dioxide concentration in the atmosphere by the end of next century. Other projections of energy production and use next century which the IPCC has developed demonstrate that alternative strategies exist for large reductions in fossil fuel emissions to be achieved.

As the WEC admits, the achievement of their scenario C (or a similar one which would lead to stabilization of carbon dioxide concentrations) will not be easy and will require both clear policies on the part of governments and cooperation from industry and from individuals. What is needed now to begin to follow that pathway, however, is neither particularly difficult or costly. It just requires clear resolve. Four areas of action are important.

- Many studies have shown that in most developed countries improvements in energy efficiency of 30 per cent or more can be achieved at little or no net cost or even at some overall saving. But industry and individuals will require not just encouragement, but modest incentives if the savings are to be realized.
- Much of the necessary technology is available for renewable energy sources (especially 'modern' biomass, wind and solar energy), which can go a long

---

Agenda 21 and energy

Agenda 21 is a document of some 400 or 500 pages accepted by the participating nations at the United Nations Conference on Environment and Development at Rio de Janeiro in June 1992. It covers a very wide range of topics concerned with the environment and development. The following paragraph 9.12 is contained in the chapter dealing with protection of the atmosphere:

Governments ... with the cooperation of the relevant United Nations bodies and, as appropriate, intergovernmental and non-governmental organizations and the private sector, should: (a) cooperate in identifying and developing economically viable, and environmentally sound energy sources to promote the availability of increased energy supplies to support sustainable development efforts, in particular in developing countries; ... (d) promote the research, development, transfer and use of technologies and practices for environmentally sound energy systems, including new and renewable energy systems, with particular attention to developing countries ...

---

way towards replacing energy from fossil fuels, to be developed and implemented. For this to be done on an adequate scale, an economic framework with appropriate incentives will need to be set up. Policy options available include the removal of subsidies, carbon or energy taxes (which recognize the environmental cost associated with the use of fossil fuels) and tradeable permits.

- Arrangements are needed to ensure that technology is available for all countries (including developing countries through technology transfer) to develop their energy plans with high efficiency and to deploy renewable energy sources (for instance local solar energy or wind generators) as widely as possible.
- With world investment in the energy industry running at around one million million US dollars per year, there is a great responsibility on both governments and industry to ensure that energy investments take long-term environmental requirements fully into account.

At the United Nations Conference on Environment and Development at Rio de Janeiro in June 1992, the countries of the world committed themselves in Agenda 21 (see box) to the action necessary to address the problems of energy and the environment. But the WEC are not optimistic that the necessary commitment exists to meet what is required, for instance to deliver their scenario C. They point out that 'the real challenge is to communicate the reality that the switch to alternative forms of supply will take many decades, and thus the realization of the need, and commencement of the appropriate action, must be *now*' (their italics)[63].

Questions **1** Estimate how much energy you use per year in your home or your apartment. How much of this comes from fossil fuels? What does it contribute to emissions of carbon dioxide?

2 Estimate how much energy your car uses per year. What does this contribute to emissions of carbon dioxide?

3 Look up estimates made at different times over the last thirty years of the size of world reserves of coal, oil and gas. What do you deduce from the trend of these estimates?

4 Estimate the annual energy saving for your country as a result of: (1) unnecessary lights in all homes being switched off; (2) all homes changing all light bulbs to low energy ones; (3) all homes being maintained 1 °C cooler during the winter.

5 Find out for your country the fuel sources which contribute to electricity supply. Suppose a typical home heated by electricity in the winter converted to gas heating, what would be the change in annual carbon dioxide emissions?

6 Find out about the cost of heat pumps and building insulation. For a typical building, compare the costs (capital and running costs) of reducing by 75 per cent the energy required to heat it by installing heat pumps or by adding to the insulation.

7 Visit a large electrical store and collate information relating to the energy consumption and the performance of domestic appliances: refrigerators, cookers, microwave ovens and washing machines. Which do you think are the most energy efficient and how do they compare with the least energy efficient? Also how well labelled were the appliances with respect to energy consumption and efficiency?

8 Consider a flat roofed house of typical size in a warm, sunny country with a flat roof incorporating 50 mm thickness of insulation (refer to Table 11.5). Estimate the extra energy which would have to be removed by air conditioning if the roof were painted black rather than white. How much would this be reduced if the insulation were increased to a thickness of 150 mm?

9 Rework the calculations of total heating required for the building considered in Table 11.5 supposing insulation 250 mm thick (the Danish standard) were installed in the cavity walls and in the roof.

10 Look up articles about the environmental and social impact of large dams. Do you consider the benefits of the power generated by hydroelectric means are worth the environmental and social damage?

11 Suppose an area of 10 square kilometres were available for use for renewable energy sources, to grow biomass, to mount PV solar cells or to mount wind generators. What criteria would determine which use would be most effective? Compare the effectiveness for each use on a typical area of your country.

12 What do you consider the most important factors which prevent the greater use of nuclear energy? How do you think their seriousness compares with the costs or damages arising from other forms of energy production?

13 In the IPCC 1995 Report, you will find more information about the LESS scenarios. In particular estimates are provided, for the different alternatives, of the amount of land which will be needed in different parts of the world for the production of energy from biomass. For your own country or region, find out how easily, on the timescale required, it is likely that this amount of land could be provided. What would be the likely consequences arising from using the land for biomass production than for other purposes?

14 In making arguments for a carbon tax would you attempt to relate it to the likely cost of damage from Global Warming (Chapter 9), or would you relate it to what is required to enable appropriate renewable energies to compete at an adequate level? From the information in Figs. 11.7, 11.12 and 11.13 and any further data to which you have access, what level of carbon tax do you consider would be likely to enable there to be greater employment of different forms of renewable energy, – (1) at the present time, (2) in 2020?

15 In discussing policy options, attention is often given to 'win-win' situations or to those with a 'double dividend' – ie situations in which, when a particular action is taken to reduce greenhouse gas emissions, additional benefits arise as a bonus. Describe examples of such situations.

16 Of the policy options listed towards the end of the chapter, which do you think could be most effective in your country?

Notes

1 *Climate Change 1995: Impacts, Adaptations and Mitigation of Climate Change,* eds. R. T. Watson, M. C. Zinyowera and R. H. Moss, CUP, 1996, Summaries, Energy Primer and chapters 19, 20, 21 and 22. See also *IPCC Technical Paper number 1* 'Technologies, Policies and Measures for Mitigating Climate Change', IPCC, Geneva, 1997.

2 1 toe = 11.7 MWh; 1 toe per day = 487 kW.

3 Adapted from G. R. Davis, 'Energy for Planet Earth', *Sci. Amer.*, **263**, September 1990, pp. 21–7.

4 *Energy for tomorrow's world – the realities, the real options and the agenda for achievement*, WEC Commission Report, World Energy Council 1993, p. 41.

5 *Energy for tomorrow's world – the realities, the real options and the agenda for achievement*, WEC Commission Report, World Energy Council, 1993.

6 A more detailed breakdown is given in IPCC 1995 'Energy Primer' It provides similar conclusions for identified resources but shows substantial differences for the more speculative estimates of ultimately recoverable resources.

7 *Energy for tomorrow's world – the realities, the real options and the agenda for achievement*, WEC Commission Report, World Energy Council, 1993.

8 *Renewable energy resources: opportunities and constraints 1990–2020*, Report 1993, World Energy Council, London.

9 *Energy for tomorrow's world – the realities, the real options and the agenda for achievement*, WEC Commission Report, World Energy Council, 1993.

10 From *Energy for tomorrow's world – the realities, the real options and the agenda for achievement*, WEC Commission Report, World Energy Council, 1993, p 27.

11 From *Energy for tomorrow's world – the realities, the real options and the agenda for achievement*, WEC Commission Report, World Energy Council, 1993, p 304.

12 From *Energy for tomorrow's world – the realities, the real options and the agenda for achievement*, WEC Commission Report, World Energy Council, 1993, p. 28.

13 From *Energy for tomorrow's world – the realities, the real options and the agenda for achievement*, WEC Commission Report, World Energy Council, 1993, p. 29.

14 H. Ishitani *et al.*, 'Energy supply mitigation options', chapter 19 in *Climate Change 1995: Impacts, Adaptations and Mitigation of Climate Change*, eds. R. T. Watson, M. C. Zinyowera and R. H. Moss, CUP, 1996.

15 From *Energy for tomorrow's world – the realities, the real options and the agenda for achievement*, WEC Commission Report, World Energy Council, 1993, p. 122.

16 From *Energy for tomorrow's world – the realities, the real options and the agenda for achievement*, WEC Commission Report, World Energy Council, 1993, p. 113.

17   From National Academy of Sciences, *Policy Implications of Greenhouse Warming*, National Academy Press, Washington DC, 1992, chapter 21.

18   From UK government estimates 1993.

19   From National Academy of Sciences, *Policy Implications of Greenhouse Warming*, National Academy Press, Washington DC, 1992, pp. 211, 212.

20   *Energy for tomorrow's world – the realities, the real options and the agenda for achievement*, WEC Commission Report, World Energy Council, 1993, p 53 (source: Society of Motor Manufacturers and Traders UK).

21   *Energy for tomorrow's world – the realities, the real options and the agenda for achievement*, WEC Commission Report, World Energy Council, 1993, p 53.

22   Example quoted in *Energy, Environment and Profits*, published by the Energy Efficiency Office of the Department of the Environment, UK 1993.

23   See for instance National Academy of Sciences, *Policy Implications of Greenhouse Warming*, National Academy Press, Washington DC, 1992, chapter 22; *Energy for tomorrow's world – the realities, the real options and the agenda for achievement*, WEC Commission Report, World Energy Council, 1993, chapter 4; and T. Kashiwagi *et al.*, 'Industry', chapter 20 in *Climate Change 1995: Impacts, Adaptations and Mitigation of Climate Change*, eds. R. T. Watson, M. C. Zinyowera and R. H. Moss, CUP, 1996.

24   Royal Commission on Environmental Pollution, 18th Report *'Transport and the Environment'*, HMSO, London, 1994.

25   Carbonaceous fuel is burnt to form carbon monoxide, CO, which then reacts with steam according to the equation $CO + H_2O = CO_2 + H_2$.

26   J. Twidell and T. Weir, *Renewable Energy Resources*, E. and F. N. Spon, 1986, p. 291.

27   *Energy for tomorrow's world – the realities, the real options and the agenda for achievement*, WEC Commission Report, World Energy Council, 1993, p 94.

28   'Large' hydro applies to schemes greater than 10 megawatts in capacity, 'small' hydro to schemes smaller than 10 megawatts.

29   Sources of comprehensive information about renewable energy are *Renewable Energy*, eds. T. B. Johansson *et al.*, Island Press, Washington DC, 1993; World Energy Council Report: *Renewable Energy Resources*, World Energy Council, London, 1993; and H. Ishitani *et al.*, 'Energy supply mitigation options', chapter 19 in *Climate Change 1995: Impacts, Adaptations and Mitigation of Climate Change*, eds. R. T. Watson, M. C. Zinyowera and R. H. Moss, CUP, 1996.

30   J. R. Moreira and A. D. Poole, 'Hydropower and its constraints' in *Renewable Energy*, eds. T. B. Johansson *et al.*, Island Press, Washington DC, 1993, chapter 2, pp.73–119.

31   Further details in *Renewable energy resources: opportunities and constraints 1990–2020*, Report 1993, World Energy Council, London, section 7.

32   J. R. Moreira and A. D. Poole, 'Hydropower and its constraints' in *Renewable Energy*, eds. T. B. Johansson *et al.*, Island Press, Washington DC, 1993, chapter 2, pp. 73–119.

33   From *Report of the Renewable Energy Advisory Group, Energy Paper Number 60*, UK Department of Trade and Industry, November 1992.

34   See Royal Commission on Environmental Pollution, 17th Report, *Incineration of Waste*, HMSO, London, 1993, pp. 43–7.

35   From *Report of the Renewable Energy Advisory Group, Energy Paper Number 60*, UK Department of Trade and Industry, November 1992.

36   D. O. Hall *et al.*, 'Biomass for energy: supply prospects', in *Renewable Energy*, eds. T. B. Johansson *et al.*, Island Press, Washington DC, 1993, pp. 593–651.

37   From *Report of the Renewable Energy Advisory Group, Energy Paper Number 60*, UK Department of Trade and Industry, November 1992, pA29.

38 See E. Mills, D. Wilson and T. Johansson in *Climate Change: Science, Impacts and Policy, Proceedings of the Second World Climate Conference*, eds. J. Jager and H. L. Ferguson, CUP, 1991, pp. 311–28.

39 From J. Twidell and A. Weir, *Renewable Energy Resources*, E. and F. N. Spon, 1986.

40 From *Report of the Renewable Energy Advisory Group, Energy Paper Number 60*, UK Department of Trade and Industry, November 1992.

41 M. Lal, 'Measures for reducing climate relevant gas emissions in India', paper presented at an Indo-German seminar IIT, Dehli, October 29–31, 1996.

42 J. Twidell and A. Weir, *Renewable Energy Resources*, E. and F. N. Spon, 1986, p. 252.

43 Adapted from J. Twidell and A.Weir, *Renewable Energy Resources*, E. and F. N. Spon, 1986, p. 100.

44 H. Ishitani *et al.*, 'Energy supply mitigation options', chapter 19 in *Climate Change 1995: Impacts, Adaptations and Mitigation of Climate Change*, eds. R. T. Watson, M. C. Zinyowera and R. H. Moss, CUP, 1996.

45 J. Twidell and C. Johnstone, 'Glasgow gains from Strathclyde's solar residences', *Sun at Work in Europe*, Vol 7 No 4, December 1992, pp. 15–17.

46 H. Kelly, 'Introduction to photovoltaic technology', in *Renewable Energy*, eds. T. B. Johansson *et al.*, Island Press, Washington DC, 1993, pp. 297–336.

47 D. E. Carlson and S. Wagner, 'Amorphous silicon photovoltaic systems' in *Renewable Energy*, eds. T. B. Johansson *et al.*, Island Press, Washington DC, 1993, pp. 403–36.

48 K. Zweibel and A. M. Barnett, 'Polycrystalline thin-film photovoltaics' in *Renewable Energy*, eds. T. B. Johansson *et al.*, Island Press, Washington DC, 1993, pp. 437–82

49 *Renewable energy resources: opportunities and constraints 1990–2020*, Report 1993, World Energy Council, London, pp. 2–49.

50 D. Lovejoy, 'Photovoltaics for rural electrification' in *Renewable Energy: Technology and the Environment*, ed. A. A. M. Sayigh, Volume 1, Pergamon Press, 1992, pp. 124–32.

51 This figure assumes that about one third of total solar energy in 2020 will be PV electricity.

52 N. Williams, K. Jacobson, H. Burris, 'Sunshine for light in the night', *Nature*, **362**, 1993, pp. 691–2.

53 *IPCC Technical Paper 3*, D. Schimel *et al.*, 'Stabilisation of atmospheric greenhouse gases: physical, biological and socio-economic implications' IPCC, Geneva, 1997, addresses this question.

54 Adapted from National Academy of Sciences, *Policy Implications of Greenhouse Warming*, National Academy Press, Washington DC, 1992, p. 355.

55 B.S. Fisher *et al.*, 'An economic assessment of policy instruments for combatting climate change, Chapter 11 in *Climate Change 1995: Economics and Social Dimensions of Climate Change*, eds. J. Bruce, Hoesung Lee and E. Haites, CUP, 1996.

56 *Renewable Energy*, eds. T. B. Johansson *et al.*, Island Press, Washington DC, 1993, p. 38; also H. Ishitani *et al.*, 'Energy supply mitigation options', chapter 19 in *Climate Change 1995: Impacts, Adaptations and Mitigation of Climate Change*, eds. R. T. Watson, M. C. Zinyowera and R. H. Moss, CUP, 1996.

57 From J. M. Ogden and J. Nitsch, 'Solar Hydrogen' in *Renewable Energy*, eds. T. B. Johansson *et al.*, Island Press, Washington DC, 1993, pp. 925–1009.

58 Adapted from J. Twidell and T. Weir, *Renewable Energy Resources*, E. and F. N. Spon, 1986, p. 399.

59 For more details of this technology see Ogden and Nitsch, 'Solar Hydrogen' in *Renewable Energy*, eds. T. B. Johansson *et al.*, Island Press, Washington DC, 1993, pp. 925–1009.

60 *Climate Change 1995: Impacts, Adaptations and Mitigation of Climate Change,*
    eds. R. T. Watson, M. C. Zinyowera and R. H. Moss, CUP, 1996, Summary for Poli-
    cymakers

61 H. Ishitani *et al.*, 'Energy supply mitigation options', chapter 19 in *Climate Change
    1995: Impacts, Adaptations and Mitigation of Climate Change*, eds. R. T. Watson,
    M. C. Zinyowera and R. H. Moss, CUP, 1996.

62 From H. Ishitani *et al.*, 'Energy supply mitigation options', chapter 19 in *Climate
    Change 1995: Impacts, Adaptations and Mitigation of Climate Change*, eds. R. T.
    Watson, M. C. Zinyowera and R. H. Moss, CUP, 1996.

63 *Energy for tomorrow's world – the realities, the real options and the agenda for
    achievement*, WEC Commission Report, World Energy Council, 1993, p 88.

12 **The Global Village**

The preceding chapters have considered the various strands of the global warming story and the action that should be taken. In this last chapter I want first to present some of the challenges of global warming, especially those which arise because of its global nature. I then want to put global warming in the context of other major global problems faced by humankind.

The challenges of global warming

We have noted in the course of our discussion that global warming is not the only environmental problem. For instance, coastal regions are liable to subsidence for other reasons; water supplies in many places are already being depleted faster than they are being replenished and agricultural land is being lost through soil erosion. Many other reasons, locally and regionally, could be listed for the occurrence of environmental degradation. However, the importance of global warming is not diminished by the existence of these other environmental problems; in fact their existence will generally exacerbate its effects – and it is often necessary to tackle all related environmental problems together.

Local degradation of the environment is generally the result of particular action in the locality. To give an example, subsidence occurs because of the over-extraction of groundwater. In these cases the community where the malpractice is occurring suffers the damage which it causes and the principle that polluters should pay the cost of their pollution is relatively easy to apply.

That the 'polluter should pay' when the pollution is global rather than local is one of the Principles (Principle 16) enshrined in the Rio Declaration of June 1992. Appropriate mechanisms are needed to apply this principle on a global scale. The particular characteristic of global warming, compared with most environmental problems, is that it is global. However, though everybody contributes to it to a greater or lesser extent, its adverse impact will not fall uniformly. Many, especially in the developing world, will experience significant damage; others will in fact gain from it. But the appreciation that an individual burning fossil fuels anywhere in the world has impact globally demands that a global attitude must be presented to the problem. In a similar way, global *benefits* accrue from all individual actions taken to reduce fossil fuel use by increased energy efficiency or by the use of renewable energy.

There is already some experience in tackling an environmental problem of global scale: the depletion of stratospheric ozone because of the injection by humans of chlorofluorocarbons (CFCs) into the atmosphere possesses similar global characteristics to the global warming problem. An effective mechanism for tackling and solving the problem of ozone depletion has been established through the Montreal Protocol. All nations contributing to the damage have agreed to phase out their emissions of harmful substances. The richer nations involved have also agreed to provide finance and technology transfer to assist developing countries to comply. A way forward for addressing global environmental problems has therefore been charted.

Moving in that direction in the case of global warming will not be easy because the problem is so much larger and because it strikes so much nearer to the core of human resources and activities – such as energy and transport – upon which our quality of life depends. However, abatement of the use of fossil fuels need not destroy or even diminish our quality of life; it should actually improve it! In tackling the problem of global warming there are particular responsibilities and challenges for different communities of expertise which generally transcend national boundaries.

- For the world's *scientists* the brief is clear: to provide better information especially about the expected climate change on the regional and local level, always keeping an appropriate emphasis on the uncertainties of prediction. Not only politicians and policymakers but also ordinary people need the information provided in the clearest possible form, in all countries and at all levels of society. Scientists also have an important role in contributing to the research necessary to underpin the technical developments, for example in the energy, transport, forestry and agriculture sectors, required by the adaptation and mitigation strategies we have described.
- In the world of *politics,* it is well over a decade since Sir Crispin Tickell drew attention to the need for international action addressing climate change[1]. Since then, a great deal of progress has been made with the signing in Rio in 1992 of the Climate Convention and with the setting up of the Sustainable Development Commission in the United Nations. The challenges presented to the politicians and decision makers by the Convention are firstly, to achieve the right balance of development against environmental concern, that is to achieve sustainable development, and secondly, to find the resolve to turn the many fine words of the Convention into genuine action.
- As I have described the likely impacts of global warming and the ways in which it can be alleviated, the role of *technology* is clear. The necessary technology is available. The challenge of its implementation, supported by appropriate investment, needs to be picked up enthusiastically and innovatively by the world's industry. Too often environmental concerns and environmental regulation are seen by industry as a threat, when in fact, they are an opportunity. As we saw in Chapter 11, the emerging technologies associated with energy efficiency in all its aspects, renewable energy production and the efficient use and recycling of materials should bring increased employment in industry at a high level of skill and technical training. Because of increasing public awareness of the environment and of the need for its preservation, the industries which are likely to grow and flourish next

century are those which have taken environmental considerations firmly on board.

- The responsibilities of *industry* (often working together with governments) must also be seen in the world context. The need to transfer appropriate technology between countries, especially in the energy sector, is particularly important. This has been specifically recognized in the Climate Convention which in Article 4, paragraph 5 states: 'The developed country Parties … shall take all practical steps to promote, facilitate and finance, as appropriate, the transfer of, or access to, environmentally sound technologies and know-how to other Parties, particularly developing country Parties, to enable them to implement the provisions of the Convention.'

- There are also new challenges for *economists:* for instance, that of adequately representing environmental costs (including those 'costs' that cannot be valued in terms of money) and the value of 'natural' capital, especially when it is of a global kind – as mentioned in Chapter 9. There is the further problem of dealing fairly with all countries. No country wants to be put at a disadvantage economically because it has taken its responsibilities with respect to global warming more seriously than others. As economic and other instruments (for instance, taxes, subsidies, regulations, standards, grants or other measures) are devised to provide the incentives for appropriate action regarding global warming by governments or by individuals, these must be seen to be both fair and effective for all nations. Economists working with politicians and decision makers need to find imaginative solutions which recognize not just environmental concerns but political realities.

Finally, it is important to recognize that the problem is not only global but long-term – the timescales of climate change and of the influence of any action on programmes such as those of forestry, energy generation or transport are of the order of several decades. The programme of action must therefore be seen as both urgent and evolving, based on the continuing scientific, technical and economic assessments. As the IPCC 1995 Report states, 'The challenge is not to find the best policy today for the next 100 years, but to select a prudent strategy and to adjust it over time in the light of new information'.[2]

Not the only global problem

Global warming is not the only global problem. There are other issues of a global scale and we need to see global warming in their context. Four problems of particular importance impact on the global warming issue.

The first is population growth. When I was born there were about 2,000 million people in the world. By the end of the century there will be close to 6,000 million. During the lifetime of my grandchildren it is likely to rise to at least 10,000 million. Most of the growth will be in developing countries; by 2020 they will contain nearly 85 per cent of the world's people. These new people will all make demands for food, energy and work to generate the means of livelihood – all with associated implications for global warming.

The second issue is that of poverty and the increasing disparity in wealth between the developed and the developing world. The gap between the rich nations and the poor nations is becoming wider. The Prince of Wales[3] has drawn attention to the strong links which exist between population growth, poverty and environmental degradation (see box).

The third global issue is that of the consumption of resources, which in many cases is contributing to the problem of global warming. Many of the resources now being used cannot be replaced, yet we are using them at an unsustainable rate. In other words, because of the rate at which we are depleting them, we are seriously affecting their use even at a modest level by future generations. Sustainable development is all about the move to a situation in which this is no longer the case.

The fourth issue is that of global security. Our traditional understanding of security is based on the concept of the sovereign state with secure borders against the outside world. But communications, industry and commerce increasingly ignore state borders, and problems like that of global warming and the other global issues we have mentioned transcend national boundaries. Security therefore also needs to take on more of a global dimension.

The impacts of climate change may well pose a threat to security. One of the most recent wars has been fought over oil. It has been suggested that wars of the future could be fought over water[4]. The threat of conflict must be greater if nations lose scarce water supplies or the means of livelihood as a result of climate change. A dangerous level of tension could easily arise, with large numbers of environmental refugees. As has been pointed out by Admiral Sir Julian Oswald[5], who has been deeply concerned with British defence policy, a broader strategy regarding security needs to be developed which considers *inter alia* environmental threats as a possible source of conflict. In addressing the appropriate action to combat such threats, it may be better overall and more

---

Poverty and population growth

The Prince of Wales, in addressing the World Commission on Environment and Development on 22 April 1992, spoke as follows[6]:

'I do not want to add to the controversy over cause and effect with respect to the Third World's problems. Suffice it to say that I don't, in all logic, see how any society can improve its lot when population growth regularly exceeds economic growth. The factors which will reduce population growth are, by now, easily identified: a standard of health care that makes family planning viable, increased female literacy, reduced infant mortality and access to clean water. Achieving them, of course, is more difficult – but perhaps two simple truths need to be writ large over the portals of every international gathering about the environment: we will not slow the birth rate until we address poverty. And we will not protect the environment until we address the issue of poverty and population growth in the same breath.'

cost-effective in security terms to allocate resources to the removal or the alleviation of the environmental threat rather than to military or other measures to deal head-on with the security problem itself.

The goal of environmental stewardship

In the western world there are many material goals: economic growth, social welfare, better transport, more leisure and so on. But for our fulfilment as human beings we desperately need not just material challenges, but challenges of a moral or spiritual kind. There are strong connections, which I drew out in Chapter 8, between our basic attitudes, including religious belief, and environmental concern. I drew a picture of humans as stewards or gardeners of the Earth. Many people in the world are already deeply involved in a host of ways in matters of environmental concern. Such concern could, however, with benefit to us all, be elevated to a higher public and political level. Al Gore, the Vice-President of the United States, has suggested (see box) that we should embrace the preservation of the Earth as our new organizing principle. The United Nations, so far as it is able, has laid out a course of action. An appropriate challenge for everybody, from individuals, communities, industries and governments through to multinationals, especially for those in the relatively affluent Western world, is to take on board thoroughly this urgent task of the environmental stewardship of our Earth.

Al Gore's five strategic goals

Al Gore, the Vice-President of the United States, has proposed a plan for saving the world's environment[7]. He has called it 'A Global Marshall Plan' paralleled after the Marshall Plan through which the United States assisted western Europe to recover and rebuild after the Second World War. Resources for the plan would need to come from the world's major wealthy countries. He has proposed five strategic goals for the plan:

- The stabilization of world population,
- The rapid creation and development of environmentally appropriate technologies,
- A comprehensive and ubiquitous change in the economic 'rules of the road' by which we measure the impact of our decisions on the environment.
- The negotiation and approval of a new generation of international agreements; agreements which must be especially sensitive to the vast differences of capability and need between developed and developing nations.
- The establishment of a cooperative plan for educating the world's citizens about our global environment.

What the individual can do

I have spelled out the responsibilities of scientists, economists, politicians and industry. There are important contributions also to be made by ordinary individuals to help to mitigate the problem of global warming. Some of these are to:

- ensure maximum energy efficiency in the home – through good insulation (see box on page 198) against cold in winter and heat in summer, and through the use of appliances with high energy efficiency (see box on page 197);
- ensure maximum energy saving, for instance by making sure that rooms are not overheated and that light is not wasted;
- support, where possible, the provision of energy from renewable sources; for instance, purchase 'green' electricity (ie electricity from renewable sources) wherever this option is available[8];
- drive a fuel-efficient car and choose means of transport which tend to minimize overall energy use – for instance; where possible, walking or cycling,
- check, when buying wood products, that they originate from a renewable source;
- through the democratic process, encourage local and national governments to deliver policies which properly take the environment into account.

Questions

1 List and describe the most important environmental problems in your country. Evaluate how each might be exacerbated under the type of climate change expected with global warming.

2 It is commonly stated that my pollution or my country's pollution is so small compared with the whole, that any contribution I or my country can make towards solving the problem is negligible. What arguments can you make to counter this attitude?

3 Speak to people you know who are involved with industry and find out their attitudes to local and global environmental concerns. What are the important arguments which persuade industry to take the environment seriously?

4 Consider Al Gore's 5 goals. Are they sufficiently comprehensive? Are there important goals which he has omitted?

5 How do you think governments can best move forward towards strategic goals for the environment such as those suggested by Al Gore? How can citizens be persuaded to contribute to government action if it involves making sacrifices, for example paying more in tax?

6 Can you add to the list in the box at the end of the chapter of contributions which the individual can make?

7 It has been suggested that the year 2000 might be declared a 'Jubilee' year in which Third World debt would be cancelled in return for appropriate environmental action. Discuss whether this is a good idea and how it might be made successful.

Notes

1 Crispin Tickell, *Climatic Change and World Affairs*, Harvard University Press, second edition, 1986.

2 IPCC *Second Assessment* Synthesis of scientific–technical information relevant to interpreting Article 2 of the UN Framework Convention on Climate Change, IPCC, Geneva, 1995, p. 17.

3 HRH the Prince of Wales in the First Brundtland Speech, 22 April 1992, published in *Threats Without Enemies*, ed. G. Prins, Earthscan, 1993, pp. 3–14.

4 The former United Nations Secretary-General, Boutros Boutros-Ghali, has said that 'the next war in the Middle East will be fought over water, not politics'.

5 Admiral Sir Julian Oswald, 'Defence and Environmental Security' in *Threats Without Enemies*, ed. G. Prins, Earthscan, 1993.

6 HRH the Prince of Wales, in the First Brundtland Speech, 22 April 1992, published in *Threats Without Enemies*, ed. G. Prins, Earthscan, 1993, pp. 3–14.

7 Expounded in the last chapter of Al Gore, *Earth in the Balance*, Houghton Mifflin Company, 1992.

8 With changes in the organization of electricity supply companies in some countries, it is becoming possible to purchase electricity, delivered by the national grid, from a particular generating source. This is already possible in Holland and by 1998 will be possible in the UK.

Glossary

Agenda 21	A document accepted by the participating nations at UNCED on a wide range of environmental and development issues for the 21st century
Albedo	The fraction of light reflected by a surface, often expressed as a percentage. Snow-covered surfaces have a high albedo level; vegetation-covered surfaces a low albedo, because of the light absorbed for *photosynthesis*
Anthropogenic effects	Effects which result from human activities such as the burning of *fossil fuels or deforestation*
Anthropic principle	A principle which relates the existence of the Universe to the existence of humans who can observe it
Atmosphere	The envelope of gases surrounding the Earth or other planets
Atmospheric pressure	The pressure of atmospheric gases on the surface of the planet. High atmopheric pressure generally leads to stable weather conditions, whereas low atmospheric pressure leads to storms such as cyclones
Atom	The smallest unit of an *element* that can take part in a chemical reaction. Composed of a nucleus which contains *protons* and *neutrons* and is surrounded by *electrons*
Atomic mass	The sum of the numbers of *protons* and *neutrons* in the nucleus of an *atom*
Biodiversity	A measure of the number of different biological species found in a particular area
Biological pump	The process whereby carbon dioxide in the *atmosphere* is dissolved in sea water where it is used for *photosynthesis* by *phytoplankton* which are eaten by *zooplankton*. The remains of these microscopic organisms sink to the ocean bed, thus removing the carbon from the *carbon cycle* for hundreds, thousands or millions of years
Biomass	The total weight of living material in a given area
Biome	A distinctive ecological system, characterized primarily by the nature of its vegetation
Biosphere	The region on land, in the oceans and in the *atmosphere* inhabited by living organisms
Business-as-usual	The scenario for future world patterns of energy consumption and *greenhouse gas* emissions which assumes that there will be no major changes in attitudes and priorities
C3, C4 plants	Groups of plants which take up *carbon dioxide* in different ways in *photosynthesis* and are hence affected to a different extent by increased atmospheric carbon dioxide. Wheat, rice and soya bean are C3 plants; maize, sugarcane and millet are C4 plants
Carbon cycle	The exchange of carbon in various chemical forms between the *atmosphere,* the land and the oceans
Carbon dioxide	One of the major greenhouse gases. Human-generated carbon dioxide is caused mainly by the burning of fossil fuels and deforestation
Carbon dioxide fertilization effect	The process whereby plants grow more rapidly under an atmosphere of increased carbon dioxide concentration. It affects *C3 plants* more than *C4 plants*
CEE	Communauté économique européenne (European Economic Community) now the EU (European Union)

Celsius	Temperature scale, sometimes known as the Centigrade scale. Its fixed points are the freezing point of water (0 °C) and the boiling point of water (100 °C)
CFCs	Chlorofluorocarbons; synthetic compounds used extensively for refrigeration and aerosol sprays until it was realized that they destroy ozone (they are also very powerful greenhouse gases) and have a very long lifetime once in the *atmosphere*. The Montreal Protocol agreement of 1987 is resulting in the scaling down of CFC production and use in industrialized countries
Chaos	A mathematical theory describing systems which are very sensitive to the way they are originally set up; small discrepancies in the initial conditions will lead to completely different outcomes when the system has been in operation for a while. For example, the motion of a pendulum when its point of suspension undergoes forced oscillation will form a particular pattern as it swings. Started from a slightly different position, it can form a completely different pattern, which could not have been predicted by studying the first one. The weather is a partly chaotic system, which means that even with perfectly accurate forecasting techniques, there will always be a limit to the length of time ahead that a useful forecast can be made
CIS	Commonwealth of Independent States (former USSR)
Climate	The average weather in a particular region
Climate sensitivity	The global average temperature rise under doubled *carbon dioxide* concentration in the *atmosphere*
Compound	A substance formed from two or more *elements* chemically combined in fixed proportions
Condensation	The process of changing state from gas to liquid
Convection	The transfer of heat within a fluid generated by a temperature difference
Coppicing	Cropping of wood by judicious pruning so that the trees are not cut down entirely and can regrow
Daisyworld	A model of biological feedback mechanisms developed by James Lovelock (see also *Gaia hypothesis*)
DC	Developing country – also Third World country
Deforestation	Cutting down forests; one of the causes of the enhanced *greenhouse effect*, not only when the wood is burned or decomposes, releasing *carbon dioxide*, but also because the trees previously took carbon dioxide from the *atmosphere* in the process of photosynthesis
Deuterium	Heavy *isotope* of hydrogen
Drylands	Areas of the world where precipitation is low and where rainfall often consists of small, erratic, short, high-intensity storms
Ecosystem	A distinct system of interdependent plants and animals, together with their physical environment
El Niño	A pattern of ocean surface temperature in the Pacific off the coast of South America, which has a large influence on world *climate*
Electron	Negatively charged component of the *atom*
Element	Any substance that cannot be separated by chemical means into two or more simpler substances
Environmental refugees	People forced to leave their homes because of environmental factors such as drought, floods, sea-level rise
Evaporation	The process of changing state from liquid to gas
FAO	The United Nations Food and Agriculture Organization
Feedbacks	Factors which tend to increase the rate of a process (positive feedbacks) or decrease it (negative feedbacks), and are themselves affected in such a way as to continue the feedback process. One example of a positive feedback is snow falling on the Earth's surface, which gives a high *albedo* level. The high level of reflected rather

than absorbed solar radiation will make the Earth's surface colder than it would otherwise have been. This will encourage more snow to fall, and so the process continues

Fossil fuels

Fuels such as coal, oil and gas made by decomposition of ancient animal and plant remains which give off *carbon dioxide* when burned

Gaia hypothesis

The idea, developed by James Lovelock, that the *biosphere* is an entity capable of keeping the planet healthy by controlling the physical and chemical environment

Geoengineering

Artificial modification of the environment to counteract *global warming*

Geothermal energy

Energy obtained by the transfer of heat to the surface of the Earth from layers deep down in the Earth's crust

Global warming

The idea that increased *greenhouse gases* cause the Earth's temperature to rise globally (see *greenhouse effect*)

Green Revolution

Development of new strains of many crops in the 1960s which increased food production dramatically

Greenhouse effect

The cause of *global warming*. Incoming *solar radiation* is transmitted by the *atmosphere* to the Earth's surface, which it warms. The energy is retransmitted as *thermal radiation*, but some of it is absorbed by *molecules* of *greenhouse gases* instead of being retransmitted out to space, thus warming the atmosphere. The name comes from the ability of greenhouse glass to transmit incoming solar radiation but retain some of the outgoing thermal radiation to warm the interior of the greenhouse. The 'natural' greenhouse effect is due to the greenhouse gases present for natural reasons, and is also observed for the neighbouring planets in the solar system. The 'enhanced' greenhouse effect is the added effect caused by the greenhouse gases present in the atmosphere due to human activities, such as the burning of *fossil fuels* and *deforestation*

Greenhouse gases

Molecules in the Earth's *atmosphere* such as *carbon dioxide* (CO_2), methane (CH_4) and CFCs which warm the atmosphere because they absorb some of the *thermal radiation* emitted from the earth's surface (see the *greenhouse effect*)

Greenhouse gas emissions

The release of *greenhouse gases* into the atmosphere, causing *global warming*

GtC

Gigatonnes of carbon (C) (1 gigatonne = 10^9 tonnes).
1 GtC = 3.7 Gt carbon dioxide

GWP

Global warming potential: the ratio of the enhanced *greenhouse effect* of any gas compared with that of *carbon dioxide*

Heat capacity

The amount of heat input required to change the temperature of a substance by 1 °C. Water has a high heat capacity so it takes a large amount of heat input to give it a small rise in temperature

Hydrological (water) cycle

The exchange of water between the *atmosphere*, the land and the oceans

Hydro-power

The use of water-power to generate electricity

IPCC

Intergovernmental Panel on Climate Change – the world scientific body assessing *global warming*

Isotopes

Different forms of an *element* with different *atomic masses*; an element is defined by the number of *protons* its nucleus contains, but the number of *neutrons* may vary, giving different isotopes. For example, the nucleus of a carbon atom contains six protons. The most common isotope of carbon is ^{12}C, with six neutrons making up an atomic mass of 12. One of the other isotopes is ^{14}C, with eight neutrons, giving an atomic mass of 14. Carbon-containing compounds such as *carbon dioxide* will contain a mixture of ^{12}C and ^{14}C isotopes. See also *deuterium, tritium*

Latent heat

The heat absorbed when a substance changes from liquid to gas (*evaporation*), for example when water evaporates from the sea surface using the sun's energy. It is given out when a substance changes from gas to liquid (*condensation*), for example when clouds are formed in the *atmosphere*

Milankovitch forcing	The imposition of regularity on climate change triggered by regular changes in distribution of *solar radiation* (see *Milankovitch theory*)
Milankovitch theory	The idea that major ice ages of the past may be linked with regular variations in the Earth's orbit around the sun, leading to varying distribution of incoming *solar radiation*
MINK	Region of the United States comprising the states of Missouri, Iowa, Nebraska and Kansas, used for a detailed climate study by the US Department of Energy
Molecule	Two or more *atoms* of one or more *elements* chemically combined in fixed proportions. For example, atoms of the elements carbon (C) and oxygen (O) are chemically bonded in the proportion one to two to make *molecules* of the compound *carbon dioxide* (CO_2). Molecules can also be formed of a single element, for example ozone (O_3)
Monsoon	Particular seasonal weather patterns in sub-tropical regions which are connected with particular periods of heavy rainfall
Neutron	A component of most atomic nuclei without electric charge, of approximately the same mass as the *proton*
OECD	Organization for Economic Cooperation and Development
Ozone hole	A region of the *atmosphere* over Antarctica where, during spring in the southern hemisphere, about half the atmospheric ozone disappears
Paleoclimatology	The reconstruction of ancient *climates* by such means as ice-core measurements. These use the ratios of different *isotopes* of oxygen in different samples taken from a deep ice 'core' to determine the temperature in the *atmosphere* when the sample *condensed* as snow in the clouds. The deeper the origin of the sample, the longer ago the snow became ice (compressed under the weight of more snowfall)
Passive solar design	The design of buildings to maximize use of *solar radiation*. A wall designed as a passive solar energy collector is called a solar wall
Photosynthesis	The series of chemical reactions by which plants take in the sun's energy, *carbon dioxide* and water vapour to form materials for growth, and give out oxygen. Anaerobic photosynthesis takes place in the absence of oxygen
Phytoplankton	Minute forms of plant life in the oceans
ppbv	parts per billion (thousand million) by volume (measurement of concentration)
ppmv	parts per million by volume (measurement of concentration)
Precautionary Principle	The principle of prevention being better than cure, applied to potential environmental degradation
Primary energy	Energy sources, such as *fossil fuels*, nuclear or wind power, which are not used directly for energy but transformed into light, useful heat, motor power and so on. For example, a coal-fired power station which generates electricity uses coal as its primary energy
Proton	A positively charged component of the atomic nucleus
PV	Photovoltaic: a solar cell often made of silicon which converts *solar radiation* into electricity
Radiation budget	The breakdown of the radiation which enters and leaves the Earth's *atmosphere*. The quantity of *solar radiation* entering the atmosphere from space on average is balanced by the *thermal radiation* leaving the Earth's surface and the atmosphere
Radiative forcing	The change in average net radiation at the top of the *troposphere* (the lower *atmosphere*) which occurs because of a change in the concentration of a *greenhouse gas* or because of some other change in the overall climate system. Cloud radiative forcing is the change in the net radiation at the top of the troposphere due to the presence of the cloud.

Renewable energy	Energy sources which are not depleted by use, for example *hydro-power, PV* solar cells, wind power and *coppicing*
Respiration	The series of chemical reactions by which plants and animals break down stored foods with the use of oxygen to give energy, *carbon dioxide* and water vapour
Sequestration	Removal and storage, for example, *carbon dioxide* taken from the *atmosphere* into plants via *photosynthesis*, or the storage of carbon dioxide in old oil or gas wells
Solar radiation	Energy from the sun
Sonde	A device sent into the *atmosphere* for instance by balloon to obtain information such as temperature and *atmospheric pressure*, and which sends back information by radio.
Stewardship	The attitude that human beings should see the Earth as a garden to be cultivated rather than a treasury to be raided. (See also *sustainable development*)
Stratosphere	The region of the region of the *atmosphere* between about 10 and 50 km altitude where the temperature increases with height and where the ozone layer is situated
Sustainable development	Development which meets the needs of the present without compromising the ability of future generations to meet their own needs
Thermal radiation	Radiation emitted by all bodies, in amounts depending on their temperature. Hot bodies emit more radiation than cold ones
Thermodynamics	The First Law of thermodynamics expresses that in any physical or chemical process energy is conserved (ie it is neither created or destroyed). The Second Law of thermodynamics states that it is not possible to construct a device which only takes heat energy from a reservoir and turns it into other forms of energy or which only delivers the heat energy to another reservoir at a different temperature. The Law further provides a formula for the maximum efficiency of a heat engine which takes heat from a cooler body and delivers it to a hotter one
Transpiration	The transfer of water from plants to the *atmosphere*
Tritium	Radioactive *isotope* of hydrogen, used to trace the spread of radioactivity in the ocean after atomic bomb tests, and hence to map ocean currents
Tropical cyclone	A storm or wind system rotating around a central area of low *atmospheric pressure* and occurring in tropical regions. They can be of great strength and are also called hurricanes and typhoons. Tornadoes are much smaller storms of similar violence
Troposphere	The region of the lower *atmosphere* up to a height of about 10 km where the temperature falls with height and where convection is the dominant process for transfer of heat in the vertical
Watt	Unit of power
UNCED	United Nations Conference on Environment and Development, held at Rio de Janeiro in June 1992, after which the United Nations Framework Convention on Climate Change was signed by 160 participating countries
UNEP	United Nations Environmental Programme – one of the bodies which set up the *IPCC*
UV	Ultra-violet radiation
WEC	World Energy Council – an international body with a broad membership of both energy users and the energy industry
Wind farm	Grouping of wind turbines for generating electric power
WMO	World Meteorological Organization – one of the bodies which set up the *IPCC*
Younger Dryas event	Cold climatic event which occurred for a period of about 1,500 years, interrupting the warming of the Earth after the last ice age (so called because it was marked by the spread of an Arctic flower, *Dryas octopetala*). It was discovered by a study of paleoclimatic data
Zooplankton	Minute forms of animal life in the ocean

Unit abbreviations

Quantity	Prefix	Symbol
10^{12}	tera	T
10^{9}	giga	G
10^{6}	mega	M
10^{3}	kilo	k
10^{2}	hecto	h
10^{-2}	centi	c
10^{-3}	milli	m
10^{-6}	micro	μ
10^{-9}	nano	n

Chemical symbols

CFCs	chlorofluorocarbons
CH_4	methane
CO	carbon monoxide
CO_2	carbon dioxide
H_2	molecular hydrogen
HCFCs	hydrochlorofluorocarbons
H_2O	water
N_2	molecular nitrogen
N_2O	nitrous oxide
NO	nitric oxide
NO_2	nitrogen dioxide
O_2	molecular oxygen
O_3	ozone
OH	hydroxyl radical
SO_2	sulphur dioxide

Index